MUTSURO SASAKI

MORPHOGENESIS OF FLUX STRUCTURE

AA PUBLICATIONS LONDON

Morphogenesis of Flux Structure
Mutsuro Sasaki

AA Publications are initiated by the Director of the AA School, Brett Steele, and produced through the AA Print Studio. aaprintstudio.net

AA Publications Editor: Pamela Johnston
AA Art Director: Zak Kyes
Design: Wayne Daly
Editorial Assistant: Clare Barrett

Typeface:
CAD (beta) designed by Zak Kyes with Grégory Ambos for this publication

Printed in England by Dexter Graphics

ISBN 978-1-902902-57-9

For a catalogue of AA Publications visit aaschool.ac.uk/publications
or email publications@aaschool.ac.uk

AA Publications
36 Bedford Square
London WC1B 3ES
T + 44 (0)20 7887 4021
F + 44 (0)20 7414 0783

The AA would like to thank Toyo Ito & Associates, Arata Isozaki & Associates and Sasaki & Partners for their help in realising this publication, and in particular Hajime Narukawa, Ayumi Isozaki, Tatsumi Terado, Takeshi Suzuki, Hirotoshi Komatsu, Motoshi Inukai and Hideaki Hamada. The AA's Shin Egashira provided valuable assistance in interviews. Translations from the Japanese are by Thomas Daniell.

Sasaki & Partners have supplied the majority of illustrations for this publication. Other photographs are reproduced courtesy of Arata Isozaki & Associates (pp. 24, 26-27, 36-39, 42-43, 52), Toyo Ito & Associates (pp. 55, 82-83, 89 top, 94 bottom, 95), Kazuyo Sejima & Ryue Nishizawa / SANAA (p. 56 bottom), Nagano Consultant Co. Ltd. (pp. 58 top, 76-77), Takenaka Corporation Co. Ltd. (pp. 71-73, 75), Hiroyasu Sakaguchi AtoZ (pp. 96-99), Japan Architect (pp. 78-79).

Dust jacket:
Island City Central Park, Gringrin, Fukuoka: construction of formwork for concrete shell

CONTENTS

PREFACE

The current focus of international structural design is the creation of new three-dimensional architectural structures that possess free, complex, mutable, fluid and organic characteristics, liberating architecture from the spell of Modernism and expanding its range. However, in order to implement these structures in a truly rational way, it is necessary to replace traditional empirically based structural design methods with mathematically based shape design methods that unify mechanics (rationality) and aesthetics (sensibility).

The shape design techniques I have been practising are the Sensitivity Analysis method and the 3D Extended Evolutionary Structural Optimisation (ESO) method. These involve using the principles of evolution and self-organisation of living creatures, adapted from an engineering standpoint, to generate rational structural shapes within a computer. I am currently experimenting with these methods to create new architectural structures. One example of their practical application is the free-curved surface shell produced by the Sensitivity Analysis method, another is the flux structure that results from the 3D Extended ESO method. The thought and method of the new morphogenesis is presented in this book through dialogues and collaborations with the architects Arata Isozaki and Toyo Ito.

Mutsuro Sasaki

While making full use of computer algorithms, Mutsuro Sasaki is now actualising the concepts of Gaudí's unfinished inverted suspension experiments. At the beginning of the twenty-first century, these previously unseen architectural structures are making their first appearance.

Arata Isozaki

The resolve to take on the challenge of the unknown can be perceived in structures designed by Mutsuro Sasaki. This tension brings a finely honed beauty to space.

Toyo Ito

MUTSURO SASAKI'S PARABOLIC SPACE

BRETT STEELE

Mutsuro Sasaki's inventive means for taming the complexities of fluid and organic form lies at the forefront of an engineering revolution in contemporary architecture. His collaborations with Arata Isozaki, Toyo Ito, Kazuyo Sejima, Ryue Nishizawa and Jun Aoki, amongst other leading Japanese architects, are yielding buildings of immense international importance to architects and engineers alike. The Sendai Mediatheque (2000, with Toyo Ito) is already a landmark for its 'diagrammatic' transposition of the structural principles concretised nearly a century ago in Le Corbusier's Maison Domino.

Sasaki's contribution at Sendai is immediately obvious when you look through the building's all-glass facade and see the almost supernaturally thin steel floor plates that define the different levels of the building. The floor plates are an extraordinary engineering (as well as construction) feat, and their minimal dimensions play an essential part in effecting the 'lightness' of the deep-plan interior. Beyond this readily visible element, however, the greater achievement at Sendai lies in Sasaki's balancing of the relationship between the homogeneous floors and the project's other significant material system, the irregular structural steel tubes that rise vertically through the building in place of a grid of simple columns. Sasaki has described how his first sight of Ito's preliminary sketch, showing the tubes and floors 'swaying like seaweed', triggered a 'subconscious response' – a sense that distortion was perhaps more 'real' than regularity. This awareness has informed all of his projects since Sendai, leading him to develop and refine a new method of structural design that produces distorted spaces and structures that melt and flow.

The shape analysis method begins by postulating from an architect's initial sketch *possible* forms as unknown quantities that are closely related but not identical to the sketch design. These are then used as a computational matrix for calculating the structural form that will achieve the assigned values for stress and deformation. Undertaken iteratively, this is a sophisticated analytic design method that gradually 'evolves' solutions by gently moulding and adjusting the elements of the overall structure in relation to one another. The approach is very much more topological than geometric in its understanding of form: alternative arrangements of members and elements are treated as formally equivalent (even though they may 'look different'), generating 'loose' connections that are allowed to fluctuate – literally change shape – until a final form is achieved. Of interest here is the way in which this kind of approach 'processes' (much more than imposes) an optimum structural solution.

Sasaki's methods produce precisely controlled and formalised solutions whose beauty lies in their underlying clarity. His engineering work can be seen not only to resonate with, but also to inform, much of the contemporary debate in architecture around surface, emergence and complexity. To a great extent, Sasaki's approach suggests an algorithmic or digital form of computational design that carries forward the pioneering work of the great engineer-architect Frei Otto in the 1960s and 1970s. Otto's analogue material models were in turn a significant advance on the famous hanging model experiments of Gaudí, whose parabolic spaces and structures are cited by Sasaki as an important inspiration for his own work.

Today's non-standard and irregular architectures pose new challenges to the empirical approaches of the architect's regular collaborator, the structural engineer. Traditionally engineers have always sought to rationalise structures in terms of the forces passing through the individual members, regardless of the overall complexity of the three-dimensional form. This rationalising imperative now faces a potentially fatal impasse: the impossibility of establishing linear relationships between components that are based on evolutionary and self-organising principles and so are multilinear or nonlinear in their arrangement. Over the past decade and a half Mutsuro Sasaki has helped to define a new direction in structural thinking, not only through his articulation of distinct analytic methods, but also by setting out a generalised set of principles for responding to the fluid, organic and above all evolutionary approaches that guide the production of contemporary architectural form.

This book includes extended dialogues between Mutsuro Sasaki and Arata Isozaki and Toyo Ito as well as a survey of four projects that illustrate the success of his collaborations with those architects. The idea of the book and the exhibition that accompanies it initially grew out of conversations with Toyo Ito, first at the architect's office in Tokyo, and then at the AA School in London, in a discussion with our students that coincided with Ito's receipt of the 2006 RIBA Gold Medal. In these conversations Toyo Ito made repeated reference to the important role played by the 'slow' structural research undertaken by his office and, more specifically, to Sasaki's collaborative input in a range of projects over the past decade. It is a period in Ito's career that has produced not only the Sendai Mediatheque but also a decided shift towards complex, curvilinear architectural form. While in Tokyo I secured a copy of *Flux Structure*, published on the occasion of Sasaki's 2005 Gallery Ma exhibition, and immediately understood the value of bringing the work to the AA to share with a larger European audience. This is an elegant engineering sensibility with immensely relevant architectural correlations and potential.

Brett Steele
Director, AA School
February 2007

A DIALOGUE WITH ARATA ISOZAKI

MUTSURO SASAKI I am very glad to have the opportunity to talk with you as part of a larger exploration of 'Original Forms of Modern Structure'. I thought we could take as our theme the 'structural genealogy' and development of contemporary architecture.

ARATA ISOZAKI It is not an easy thing to define the 'origins' of modern structure, but we might start by considering Antonio Gaudí, a particular interest of mine. Gaudí has been in the public eye lately, as 2002 was the 150th anniversary of his birth. Whenever a journalist asks me my opinion of him I always say, 'I hate Gaudí', though I invariably qualify this by adding, 'It is the interpretation of Gaudí by Japanese people that I hate'. The fashion for referencing Gaudí's shapes is also meaningless. Personally, I think the most important thing to understand is that Gaudí looked for his own structural methods. He genuinely thought about structure, incorporating the traditional technique of the Catalan vault, for instance, into his architecture. Gaudí had a clear direction for his architectural composition, which makes him stand out among the many architects of Art Nouveau and Modernismo.

1.

Recently, various extrapolations of Gaudí-like structural concepts have begun to appear, based on elements such as conoid curves. Though similar things were attempted before, they lacked a substantial structural basis.

SASAKI I think Gaudí was the first architect to be concerned with parabolas, which up to then had been mainly used by engineers to manufacture things in a structurally rational way. Gaudí looked at the parabola as a catenary curve, and right from the start applied it to many buildings. I think this eventually led to the three-dimensional hanging model experiments.

If we consider the work of modern masters such as Mies van der Rohe and Le Corbusier, we can see that the basis of modern structure is essentially industrial, founded on the mass-produced materials of steel and concrete. However, I would place Gaudí in the pre-modernist category of 'stone'. His work is basically an architecture of stone and brick, although these materials were ultimately replaced with reinforced and precast concrete.

2.

1. Temple Expiatori de la Sagrada Familia, Barcelona
2. Sagrada Familia School, Barcelona

ISOZAKI In a sense, he was also using bricks as a substitute for stone.

SASAKI That's right. Although Gaudí had an extremely good engineering sensibility, he dealt with stone and brick, which were readily available in Catalonia, and cheap. In terms of materials, Mies and Le Corbusier went with the flow of industrialisation in a comparatively obedient way. Mies fabricated architecture from glass and steel, Le Corbusier used concrete. However, I somehow think the schemes of these three individuals have their own distinct origins.

ISOZAKI I agree.

SASAKI We might suppose that the Parthenon is the origin of architecture throughout Europe, as it is representative of Greek architecture and philosophy. It seems to crop up all the time, not just in architecture but in every field. Broadly applying this concept of origin to Modernism, we could say that Mies exhibited a typically Platonic type of logic, whereas Gaudí's work had an Aristotelian complexity. In this sense, their approaches cannot be grouped within the same category of Modernism.

ISOZAKI My interpretation is close to yours. However the brick prototype linked to Gaudí was first developed in Rome. Unlike the Greek frame – a trabeated structure using a linear element as a lintel – the arch and the dome are assembled by combining components made of processed fragments. A dome may be composed of bricks only, as we know from looking at the Parthenon. A dome spontaneously establishes an internal space, but reduces the perimeter to merely a wall – the pediments and columns adopted from Greece no longer define an enclosure, but are integrated with the facade. From a historical viewpoint, Rome referred to Greece, but the structural use of brick gave rise to a distinctly Roman conception that is linked via the Romanesque to the Gothic, and hence to stone. Rome was the initial model for the Renaissance, but looking back we ultimately arrive at Greece. There may still have been a link between these models and modern materials, such as steel and concrete, when they first appeared in the nineteenth century, though there has been a major parting of ways since then. Seen from this angle, Mies is instantly linked to Greece. He is perhaps in the lineage of Schinkel. Le Corbusier, on the other hand, was interested in both the curved lines of arches or domes and linear frame assemblages such as the

Domino. So Le Corbusier is unconsciously joined with the orthodox flow of the Renaissance, exemplified by Brunelleschi. Of course there are differences between Mies and Le Corbusier that can be attributed to their respective national characters, but as architects thinking about architecture, we can say that one thought in terms of linear materials extending horizontally and vertically, whereas the other conceived his work in terms of continuous surfaces such as roofs and masses.

SASAKI Certainly, I feel that there is a very fundamental difference in conception between Mies and Le Corbusier. In your own works, you use Corbusian surfaces as well as 'Miesian' lines.

ISOZAKI I go back and forth between them. At least, I did so prior to the appearance of today's new structural principles.

SASAKI In terms of global trends, we've seen the emergence of complex, freeform architecture, illustrated by the Bilbao Guggenheim (1999),

3.

Sendai Mediatheque (2000) and Yokohama International Port Terminal (2002). What are your thoughts on these three works of contemporary architecture? I guess they offer pointers for thinking about the future of structure.

ISOZAKI In one respect, Bilbao is 'freehand architecture'. To make it, freehand drawings were sent to the West Coast of the US to be processed by software that substituted mathematical lines for the initial rough sketches, converting them into concrete shapes. Allocating mathematical formulas to the individual curved lines allows them to be used in preliminary calculations, but these calculated shapes cannot currently be assembled as surfaces, so they are disassembled into linear steel frames that follow three-dimensionally curved surfaces. In the case of Bilbao, the curving titanium was applied just to the outer surfaces. At the rear of the

4.
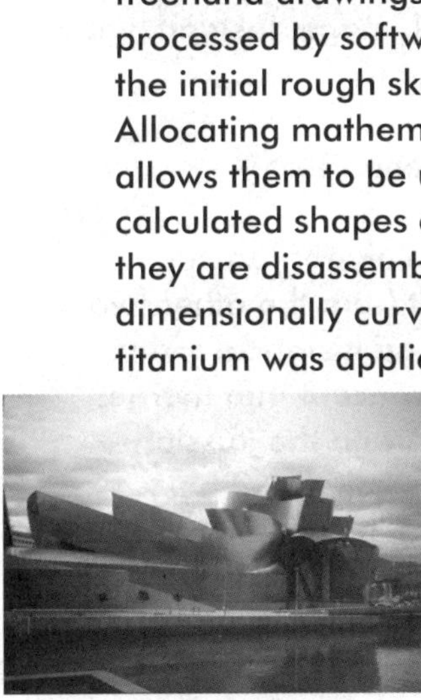

building, the steel frames remain in full view. Considered from a Miesian viewpoint, these details are not fully resolved. I would describe them as 'poor details', though in Gehry's aesthetic there is a certain merit in substituting this poor condition for conventional beauty. As a consequence, the work is not at the level of

3. Yokohama International Port Terminal
4. Guggenheim Museum, Bilbao

structural innovation – it is a freehand sculpture that happens to have galleries and toilets inserted in it, so it may be understood as architecture. It is acceptable to make something like this if you have Gehry's degree of confidence in a 'poor' aesthetic. However, if it is just mimicry, everything becomes poor.

SASAKI And with regard to Yokohama?

ISOZAKI I was head of the competition jury, so I was aware of the process right from the start. This was not a work of architecture achieved with a large-scale, comprehensive morphological theory. Neither Alejandro Zaera-Polo nor Farshid Moussavi understood how to implement the structure before they went to discuss it with Kunio Watanabe. Only the shape had tentatively been established. Moreover, this shape was defined by a very primitive numerical formula. The initial idea was to insert a honeycomb between a double slab composed of two steel plates. In other words, the proposal treated everything as a single smooth surface. But for all the talk about a honeycomb, there was a truss inside by the time the project was handed over to Watanabe, and then everything was replaced with trusses. The initial seamless flat surface became a hidden shadow – the folded-plate structure became something we have all seen before. So, if this is the simple realisation of an overall system and shape first made with a computer, I still have doubts about the process. Nonetheless, I find the roughness of the bare Corten steel and the scale of the continuous surfaces to be superbly interesting. With structural analysis, these things should have developed into a smooth, streamlined overall image. Instead, what was implemented was an origami-like condition.

5.

SASAKI And what about Sendai?

ISOZAKI I think Sendai has an approach that's completely different from the other two projects. Looked at in broad historical terms, it may be understood as a concept that merges the Domino system (1916) with the Russian Constructivist Shukhov Radio Tower (1922). I think that it has inherited the original intentions of Modernism. It is assembled extremely well, and also has an agreeable sense of scale.

A balance of scale has been skilfully achieved by implementing the flat slabs in a different way from Yokohama.

The details were thoroughly examined and the architecture appears extremely gentle. The glazing details around the periphery are far better resolved than in the two buildings previously mentioned. Rather than poor, this could be called rich detailing.

SASAKI With regard to form, these three works of architecture make lavish use of nonlinear curved surfaces employing mathematical formulas. However, it's just their shapes that are nonlinear, and essentially there is no relation between the nonlinear analysis and the generation of their structural forms. Your book *Demiurgomorphism: The Creator's Tenets* touches on undecidability and ethics. Recently, by uniting with computer graphics and artificial life theories, architectural design has been shifting from abstract geometrical forms to biomorphic organic forms. Is this applicable to projects such as Bilbao?

ISOZAKI Of course it's applicable, but my intention was to indicate tendencies among slightly younger architects. People such as Greg Lynn or Karl Chu epitomise a generation that has begun to conceive architecture using only computers. They are trying very hard to make architecture through organic animations. For instance, they might gradually transform a roof placed on a square box while also making the walls bulge and contract. In this way, a complex form can arise from a conventional cube. Teaching this has become the dominant trend at several universities in the United States.

SASAKI I think these manipulations remain thoroughly self-indulgent and arbitrary. It could hardly be said that the work is based on the fundamental principles of biotechnology. Yet the same programs could be used to carry out, say, a simulation of the flowering of a plant, or the sequential transformation of the appearance of an animal's internal organs. Whether or not this can be brought into architecture is another question. The shapes produced by simulations are certainly interesting, but the point is: 'why do them?'

ISOZAKI At this stage, we can just say the shapes are interesting. However, sooner or later it will be a problem if no one comes up with a sound basis for them. Another problem is how the shapes are to be materialised. In many cases these people are hired as consultants by other design offices. With the plan roughly decided, they input the initial shape into the program and then make proposals, 'How about this shape?' Taking the roofs and walls designed using a conventional plan, they transform them into peculiar shapes and then return them to the architect. Although buildings have been realised using this method, they're generally not the kind that get featured in magazines, because they're just slightly curved panels attached to very ordinary frames. I can only describe these people as having very meagre abilities as architects.

I think perhaps there are limits to the method of manipulating forms itself. It's fun to watch individual volumes being organically transformed on a computer screen, but these forms are purposeless. So at events like the ANY conference series, I always ask a very simple question: 'At what point do you decide to stop?' That is to say, what is the logic behind the formal decisions?

SASAKI The shapes are baseless – they just keep moving.

ISOZAKI The other day, at a symposium about architecture and the city, I had a conversation with the Italian philosopher Massimo Cacciari. I asked him how we should construe a complete image of the city in our present situation. The conventional way of thinking was to assemble this complete image piece by piece. There's no point to that now. Even if we have a complete image in mind, it soon becomes outdated and we are forced to make up another complete image. In this context, a clear direction is necessary. This is a question I also raised in my book *Toshidezain no Houhou (Methods of Urban Design)* (1962). Previously, proposing the complete image was the job of the architect. But it is clear that there's no point in trying to propose a complete image for the city, whether it's a decade or a century from now. The appearance of the future changes from moment to moment. Despite this we still seem to think that determining a complete image also means determining the details. This is a delusion.

SASAKI It is more realistic to expand from local rules without top-down control. If the whole is entirely generated from a set of local rules, it will not be subject to a single overriding constraint. As an example, spline functions were used at the Beijing National Grand Theatre (1998). The merit of this mathematical function is that even if a local portion of a curved surface is amended, it does not feed back to the whole.

ISOZAKI I see what you mean.

SASAKI Life works in exactly the same way. Even with your own body, the whole is made from local rules. As you have mentioned, cities like Tokyo and Shanghai can be viewed as unreal worlds lacking in complete images. But we can reverse-analyse how they grow into a complex whole through the bottom-up actuation of each part with local area rules. With your recent Kitagata Community Centre (2002–06), I tried to use this extremely primitive method to make a proposal in the field of structural mechanics.

ISOZAKI Even thinking that this conception could be realised seems like a historic step forward. From the nineteenth century on, the modern theory was to conceive of a complete image, then move directly towards it. A new approach will naturally require a completely different way of doing things. In this context, I think it is a problem that the concept of planning, as in planned economies or city planning, has all but disappeared. Even if we have intellectually understood that planning concepts have become ineffective, we have not come up with any viable alternatives. Perhaps that's because the conception has not been overturned from the base. We need first of all to shelve the modern theory and extract a complete image through trial and error on local areas. I think there's an important logic to this process. However, when developing this new logic, a few criteria are required to determine the smaller parts without postulating a complete image. But I can't yet imagine how… Perhaps these criteria are something hitherto unseen, unimagined. Well, I have no option but to start with such arbitrary decisions. The reason for my somewhat critical series of writings about the current biomorphic approach is that, while using software that can evolve local areas, they seem only to be looking at the complete image. Methodologically, this is completely different from what we are thinking about.

SASAKI In this sense, I am convinced the projects you are currently working on suggest a new direction for you. The first test of this new methodology is the large roof of Kitagata. I'll try to give a straightforward explanation of the structure, although it may become a bit specialised. As we've already mentioned, basic concepts of artificial life are premised on systems that operate in a bottom-up way, without any top-down rules determining the whole. Following this approach, I tried

6.

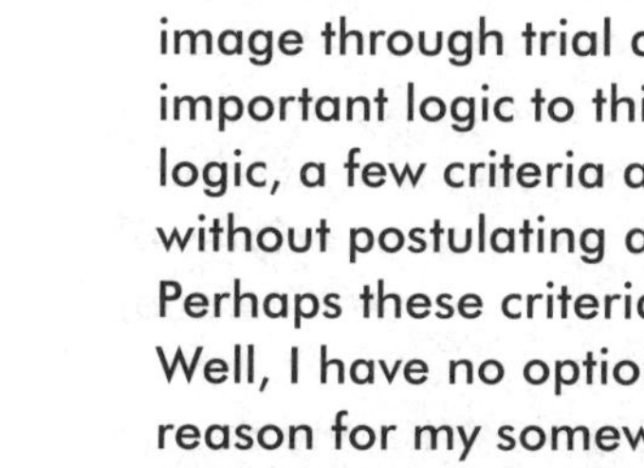

6. Roof shape study for Kitagata Community Centre

first of all to reevaluate the basic principles of artificial life, such as individual self-organisation and evolution from a seed, from an engineering viewpoint. This formed the basis for a shape analysis method (Evolutionary Structural Optimisation) that theoretically creates an optimum structural form able to adapt to a mechanical environment such as gravity, while following the basic principles of mechanics.

In ordinary structural analysis, the structural frame is assigned some kind of shape with predetermined coordinates, and then it follows the process of calculating the deformation and stress. By contrast, in shape analysis, the target values for stress and deformation are used to define the most desirable structural form. It could be called inverse analysis, in that it is the opposite of ordinary analysis. In this instance we used the Sensitivity Analysis method, which is one of the shape analysis methods being developed in my university laboratory. Seeking the optimum structural form of a free-curved surface shell, we take an unknown quantity of design parameters as coordinates in the Z axis, then try to minimise the objective function of the strain energy C (Z,U(z)) under a variety of design criteria. Individual elements evolve according to local rules in response to various changes in the design criteria. As a result, the overall structural form also undergoes a variety of transformations, theoretically producing a mysterious, hitherto unseen form that is in accordance with the principles of structural mechanics. This is pursued through nonlinear iterations within a computer.

With regard to establishing initial values for the design criteria, I think the discussion about ethics and decidability may have a bearing on this kind of theoretical method. My intention is to provide some kind of logic for structural aesthetics and ethics that may become a basis for determining form. I think that it is at last possible to provide a solution, albeit a partial one.

ISOZAKI By ethics, we do not mean a moral system but rather some kind of common standard by which the logic can be carefully evaluated and generally understood. Without this, there is no way to grasp the invariably scattered elements. Logic is a tool for pursuing, at most, mathematical equations. The architecture that we are now able to assemble should transcend logic, whether as aesthetics or ethics. Underlying the conception of the single large roof in the Barcelona (Blanes) project is the idea of making something uncertain within a situation of uncertainty. I would be pleased if this idea were incorporated and reflected in discussions at the level of architectural discourse, as well as at the level of structure. Having resolved this, other problems would then appear. I had the feeling that if all these indefinite elements were superimposed, it might unexpectedly become an interesting project.

We can largely understand the characteristics and defects of the three buildings previously discussed, including Bilbao, in this light. It would be good if we could conceive a method that surpasses them. The structure for Kitagata employs similar continuous curved surfaces to the Yokohama Ferry Terminal, but creates a mysterious impression in contrast to the purely formal manipulations of Yokohama, which disregard material characteristics.

SASAKI The difference is that the form and the structure of Yokohama are not generated from local areas.

ISOZAKI That's exactly right. Interesting shapes can be found using computer programs. A sense of dynamism, of movement, can be produced. However, that's the only feeling the spaces establish.

We're aspiring to a condition that will provide not a single solution for defining architecture but a range of solutions. The key to this appears to be tolerance, which allows the lines of a building to be slightly sloping instead of straight, without it collapsing. If this slight tolerance is only conspicuous in one part, for example, a straight line may be sloped, or twisted, or may even appear to be on the verge of breaking. Gaudí optimised this. What I feel when I look at Gaudí is the sense of a structure that incorporates extremely subtle levels of tolerance, producing the maximum visual illusion.

SASAKI In the case of Gaudí, the angles of the columns were determined by model experiments that subtly manipulated the axial forces through the inner cores of their cross sections. But to be honest, the question of why the tubes of Sendai are inclined at particular angles can't be answered on the basis of structural logic alone. Although they aren't structurally irrational, they can't be called positively rational either.

ISOZAKI In a sense, the architect and the designer must escape from the pure engineer and absorb and assemble many indefinite factors. Many more design possibilities are opened up this way.

There are many somewhat 'wrinkled' expressions to be found in Gaudí, but the excess parts are applied to the framing, where the simple logic of the Catalan vault is manifest. In comparison, Bilbao is only excess. It is boneless, comprising only excess parts. Conversely, Sendai has subtle slippages and subtle differences.

SASAKI Of course, taking these subtle differences into account structurally gives an adequately rational basis for the inclined tube columns. However, the columns give priority to the architect's design, not to structural logic.

ISOZAKI You mean it is the structural engineer who guarantees that the design can be modified? Without this guarantee, we architects would have considerable problems.

SASAKI In fact, reflecting on the lack of structural logic in the inclination of the tubes has led me to begin grappling with new research themes. Another new shape analysis method called 3D Extended Evolutionary Structural Optimisation (ESO) is now being developed in my university laboratory. I think it could be applicable to your Florence Station project.

7.
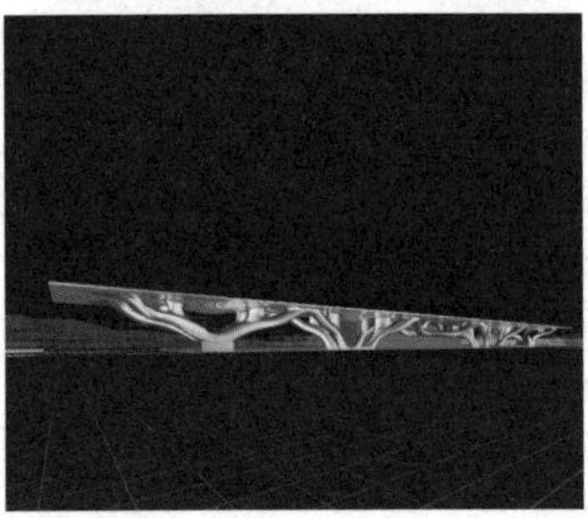

This is an edited version of a dialogue first published in Japanese in GA JAPAN 57, July/August 2002, as part of an exploration by Mutsuro Sasaki of 'Original Forms of Modern Structure'.

7. Florence New Station

PROJECTS WITH ARATA ISOZAKI

FLORENCE NEW STATION

2002

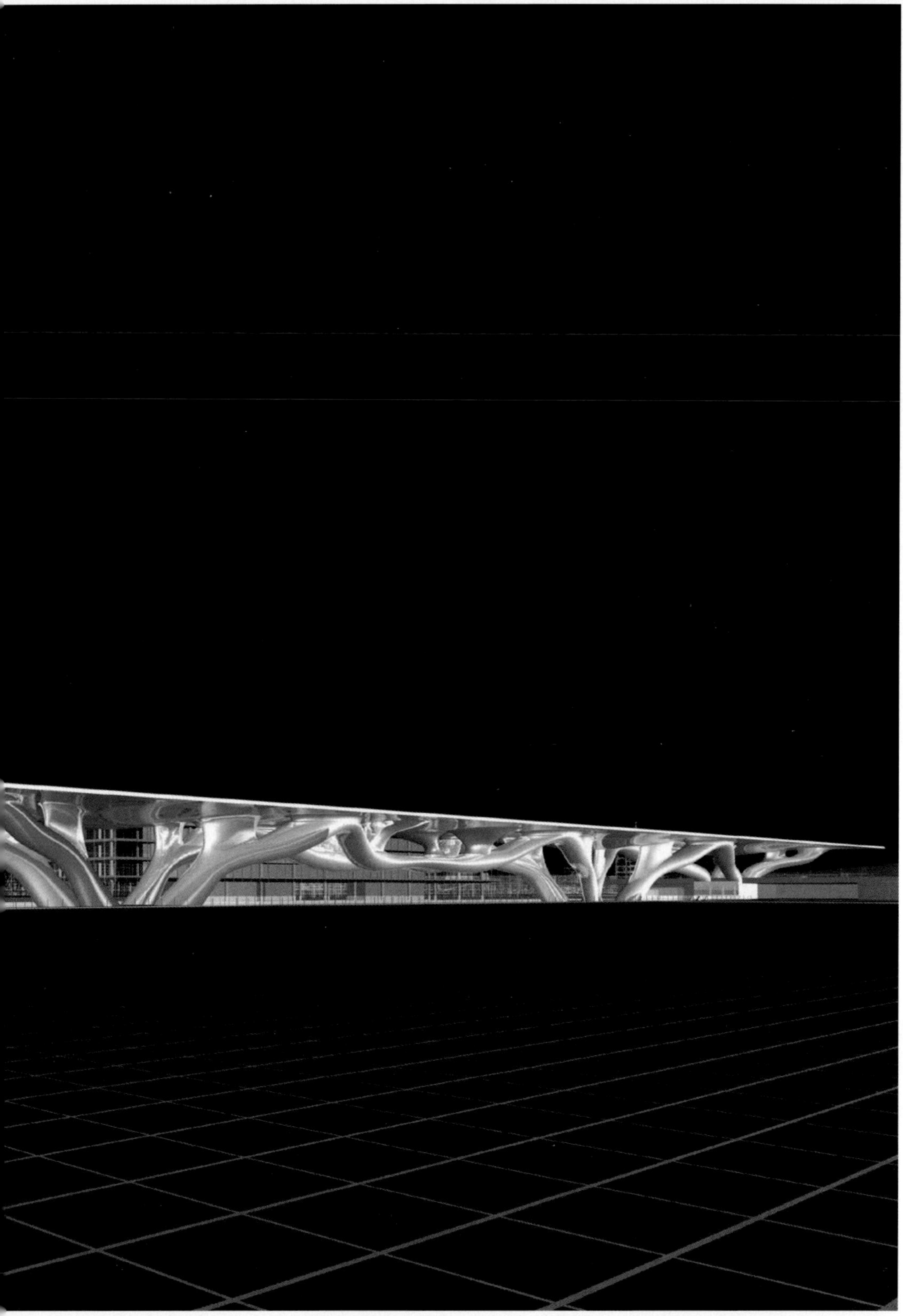

This was an entry in an international competition to design a new station in Florence, Italy. Using shape design (Extended Evolutionary Structural Optimisation (ESO) method), an enormous flux structure was created, 400 metres long by 40 metres wide by 20 metres high. The structural elements were optimally formed within a three-dimensional space while satisfying the given design conditions, and the resulting structural shape manifests maximum mechanical efficiency with a minimum use of materials. The architectural constraints during the evolution of the design were to keep the upper surface of the rooftop constantly flat and the bottom surface at the centre 12 metres or less from the roof.

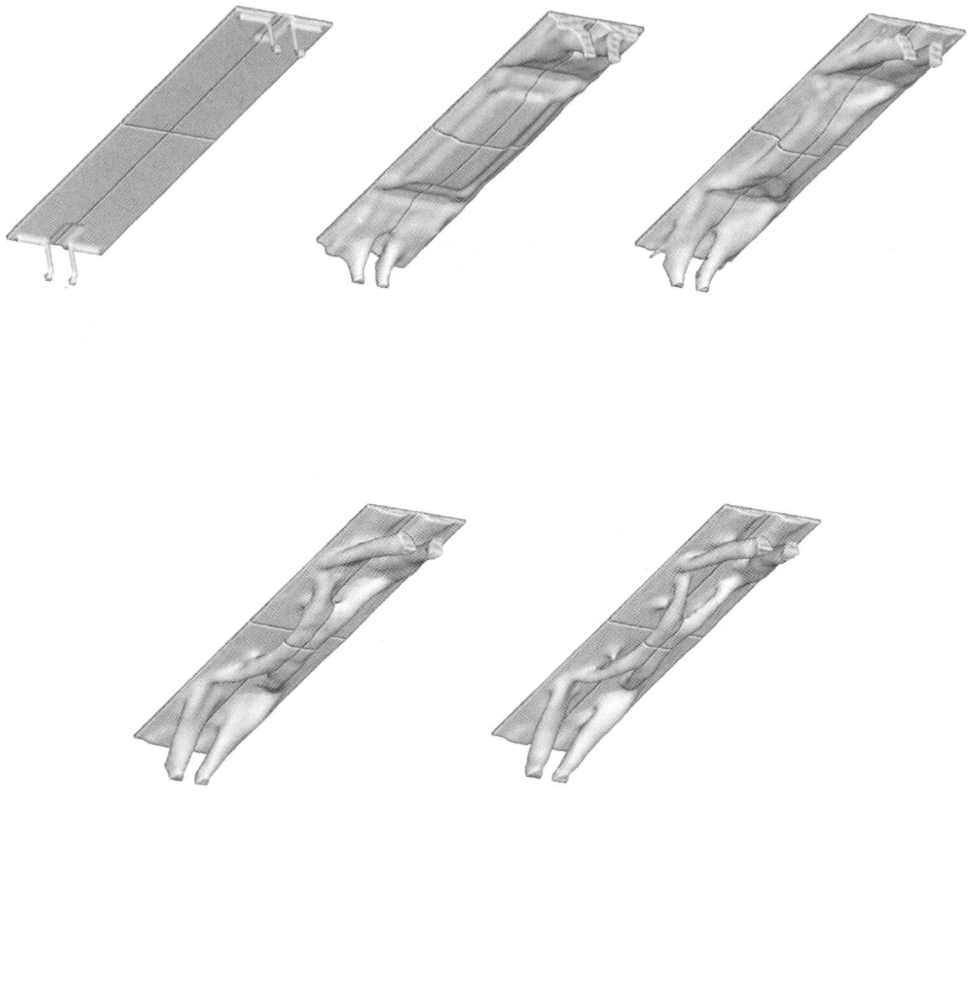

Process of shape analysis by
3D Extended ESO method

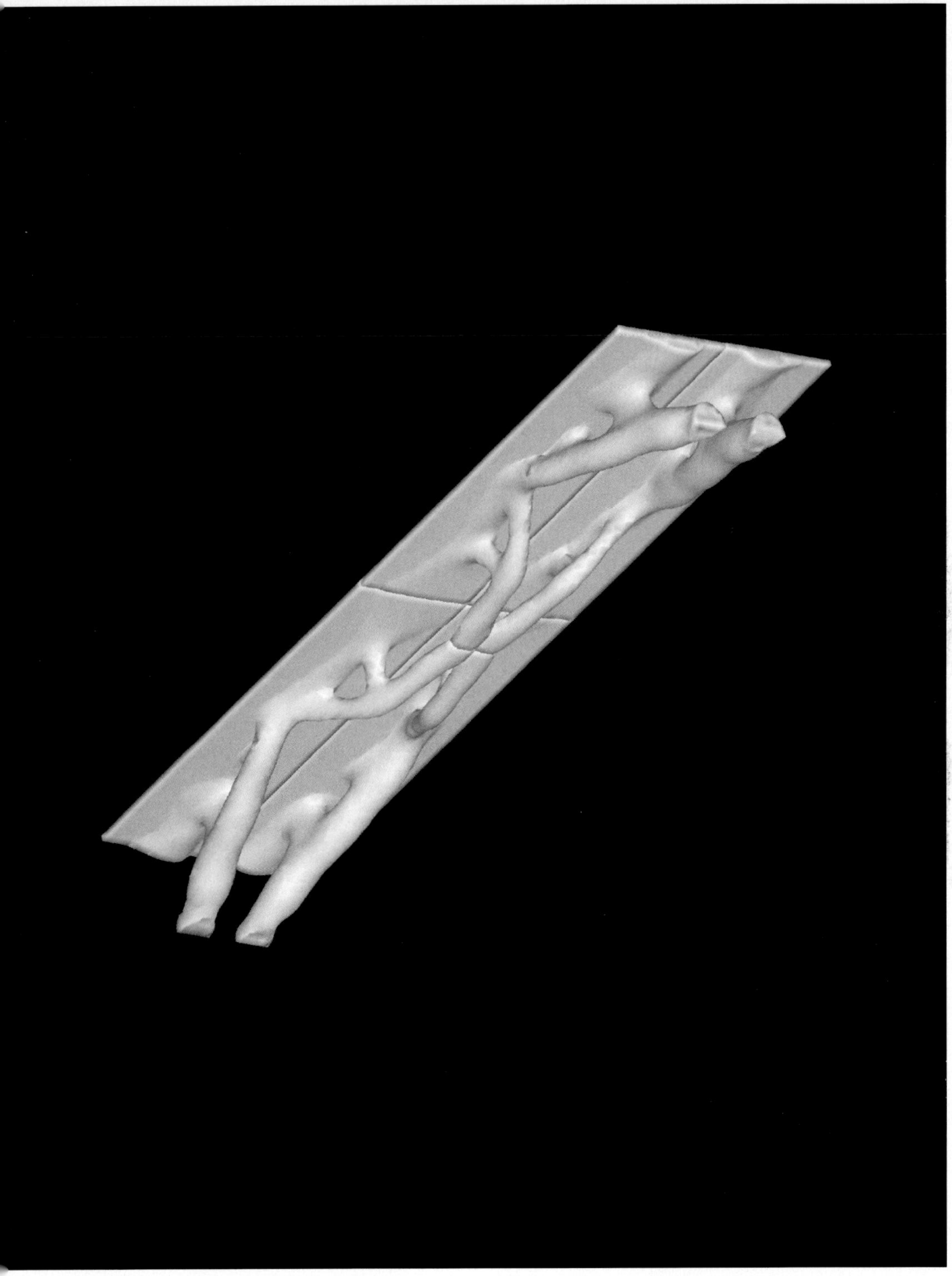

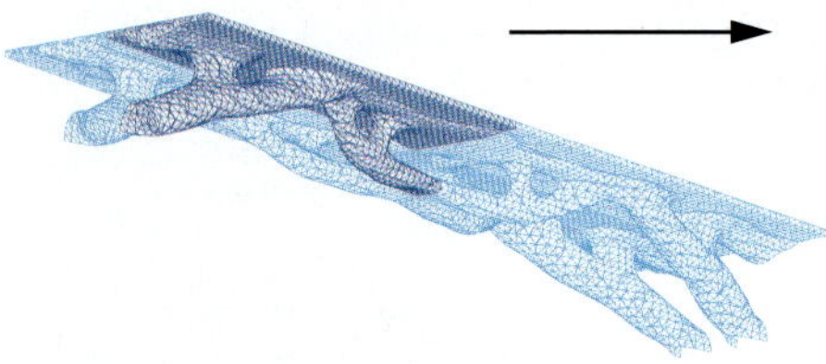

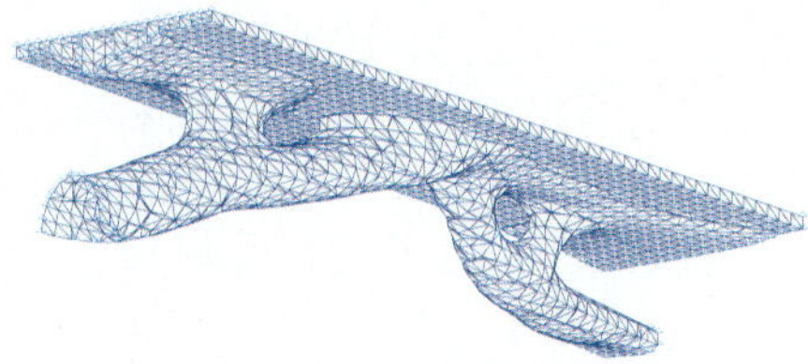

Full model

Analysis model

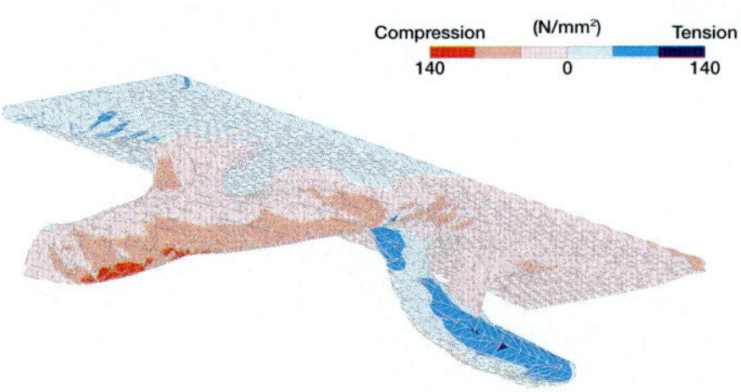

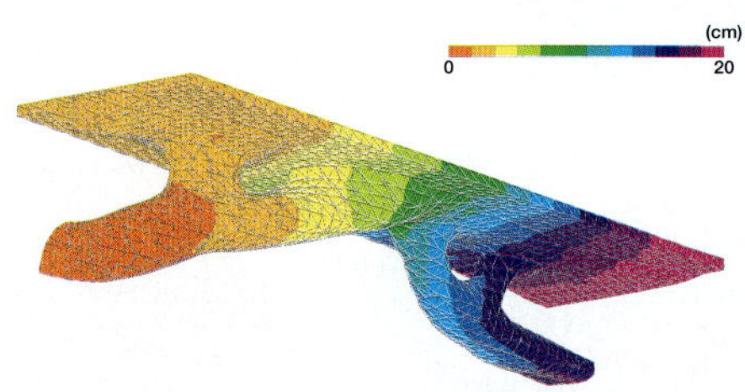

Distribution of principal stress

Distribution of displacement

FEM analysis of structure

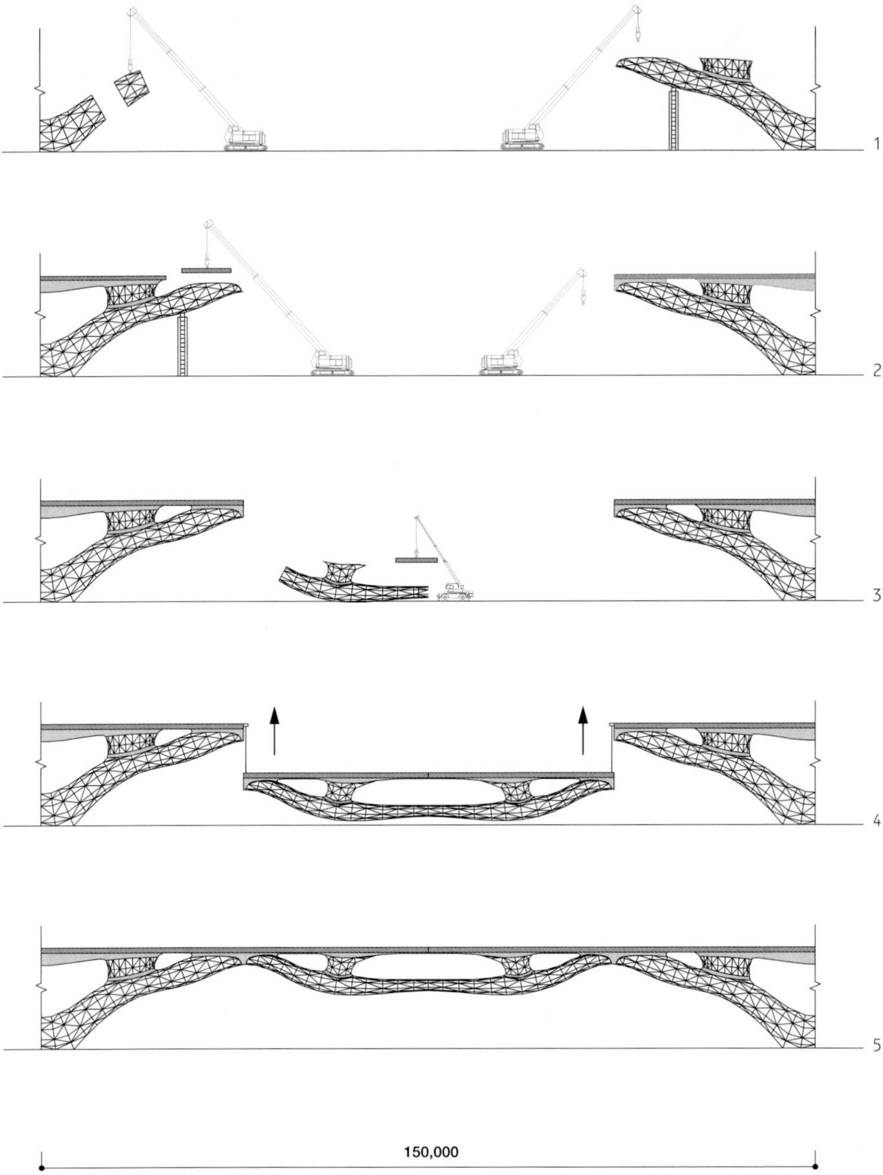

Sequence of assembly
1. The prefabricated units for the 'arch-tube' are transported to the site: the size of each unit is determined by the load-bearing capacity of the trailer or other form of transport
2. Construction of cantilevered truss – the sandwich roof plates are set on the arch-tube
3. The central part – consisting of the catenary tube and sandwich roof plate – is pre-assembled on the ground
4. The central part is raised and connected to both sides
5. Completed construction

Structural diagram

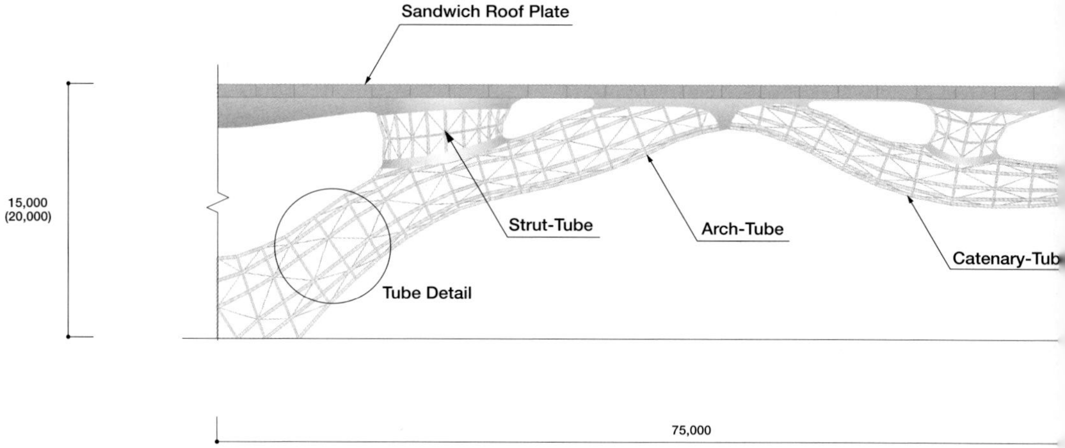

Roof detail

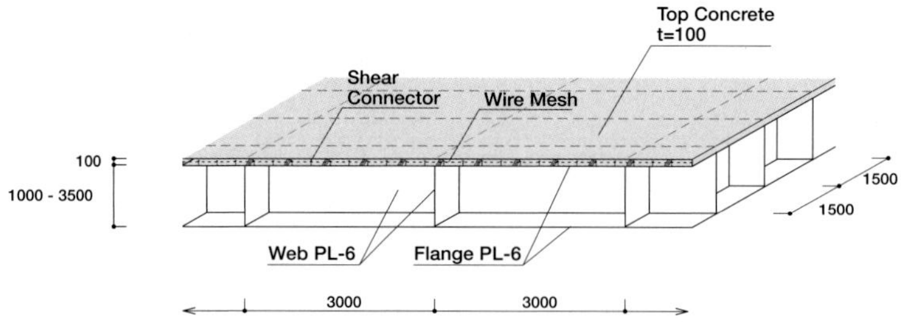

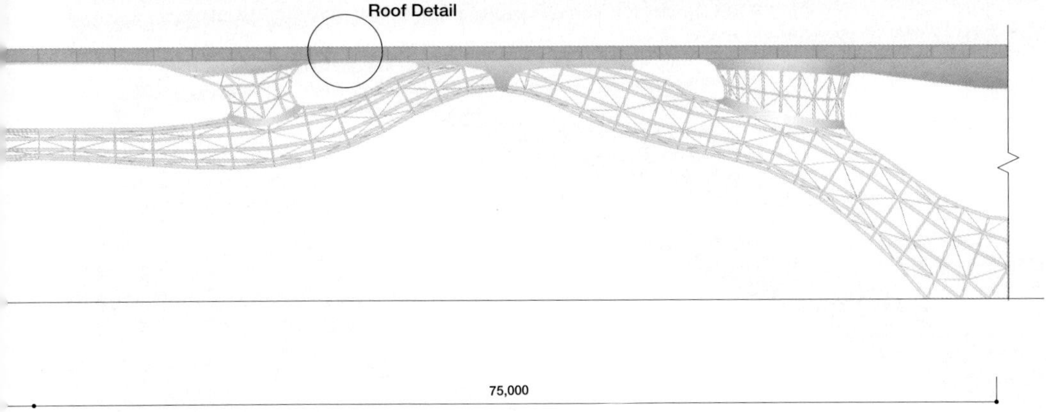
Roof Detail
75,000

Tube detail

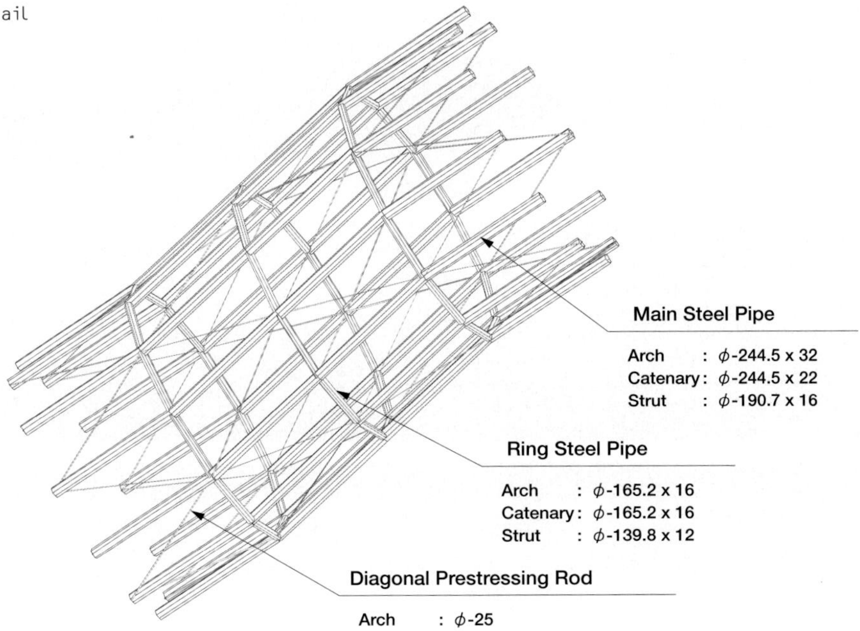
Main Steel Pipe
Arch : ϕ-244.5 x 32
Catenary: ϕ-244.5 x 22
Strut : ϕ-190.7 x 16
Ring Steel Pipe
Arch : ϕ-165.2 x 16
Catenary: ϕ-165.2 x 16
Strut : ϕ-139.8 x 12
Diagonal Prestressing Rod
Arch : ϕ-25
Catenary : ϕ-25
Strut : ϕ-25

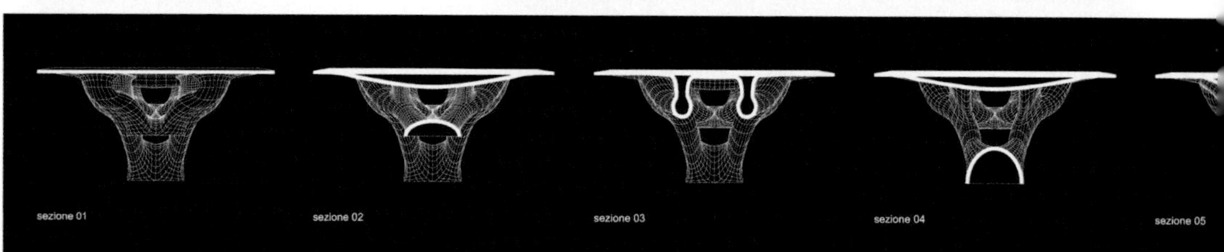

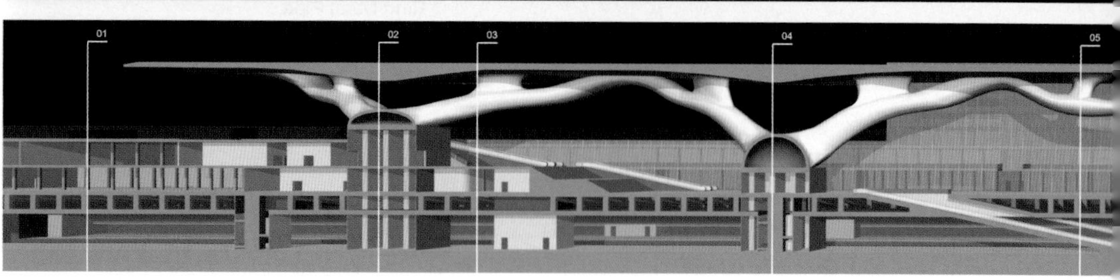

Top: Roof structure from above
and within the station hall
Centre: Cross section
Bottom: Long section through station hall

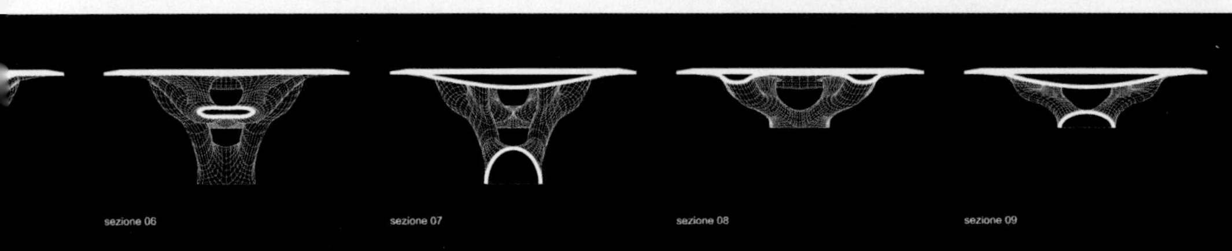
sezione 06
sezione 07
sezione 08
sezione 09

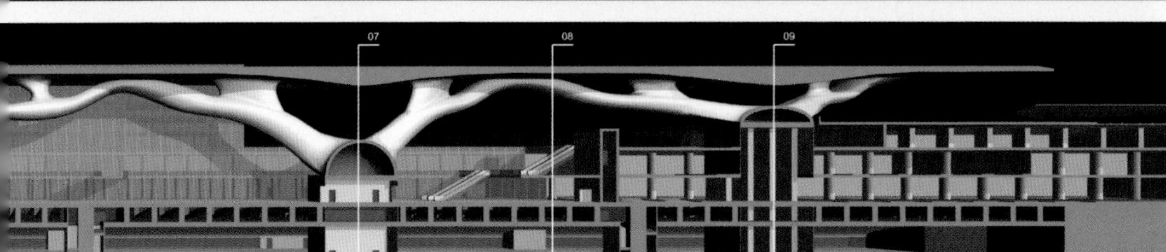
07
08
09

PROJECTS WITH ARATA ISOZAKI

QATAR EDUCATION CITY CONVENTION CENTRE

2003

This is a design for the entrance to a convention centre in Doha, Qatar. Using shape design (Extended Evolutionary Structural Optimisation (ESO) method), an enormous flux structure was created, 250 metres long by 30 metres wide by 20 metres high. The structural elements were optimally formed within a three-dimensional space while satisfying the given design conditions, and the resulting structural shape manifests maximum mechanical efficiency with a minimum use of materials. The architectural constraints during the evolution of the design were to keep the upper surface of the rooftop constantly flat and to place along its length two supporting points with a span of 100 metres.

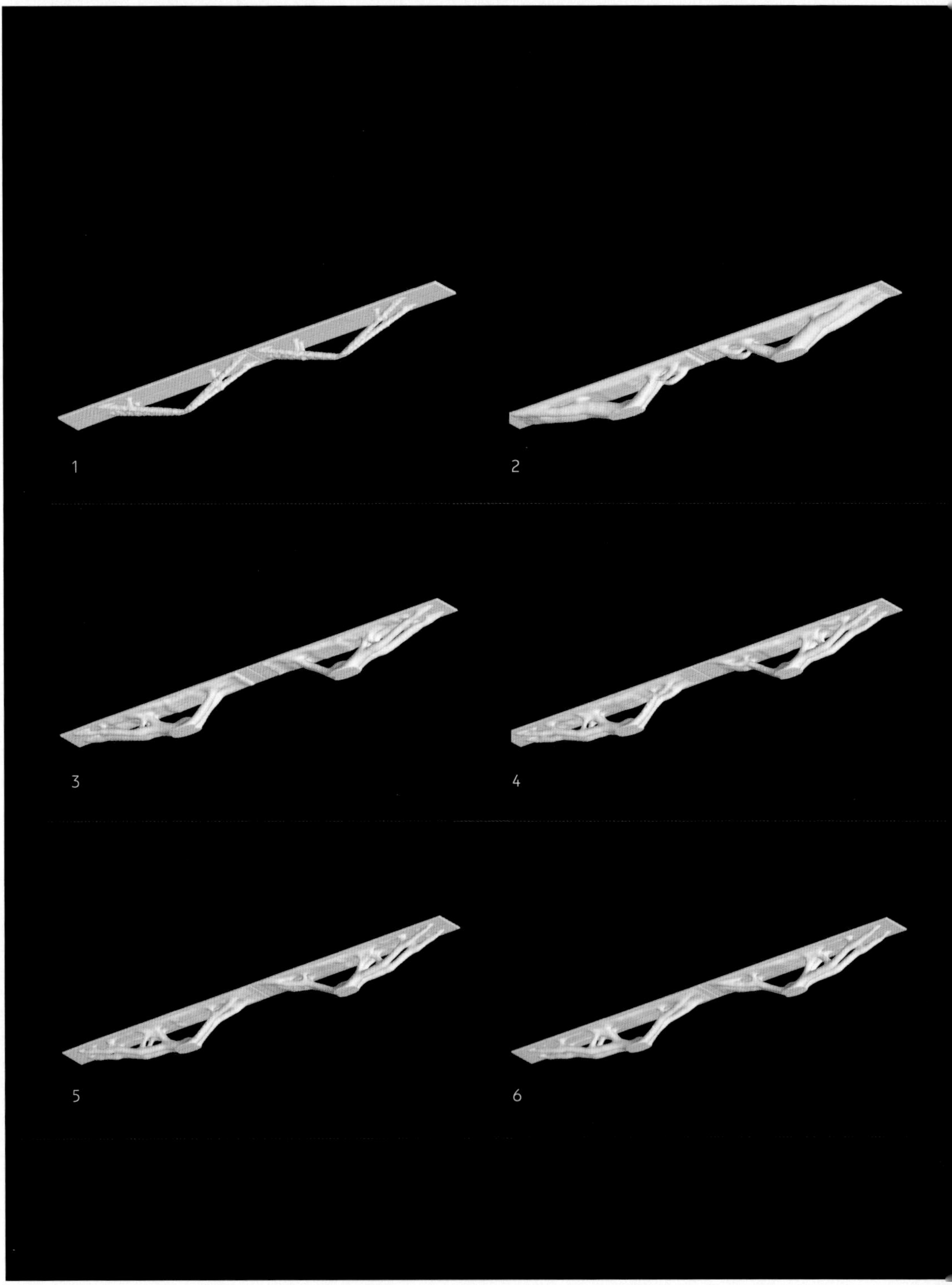

Process of shape analysis by 3D
Extended ESO method

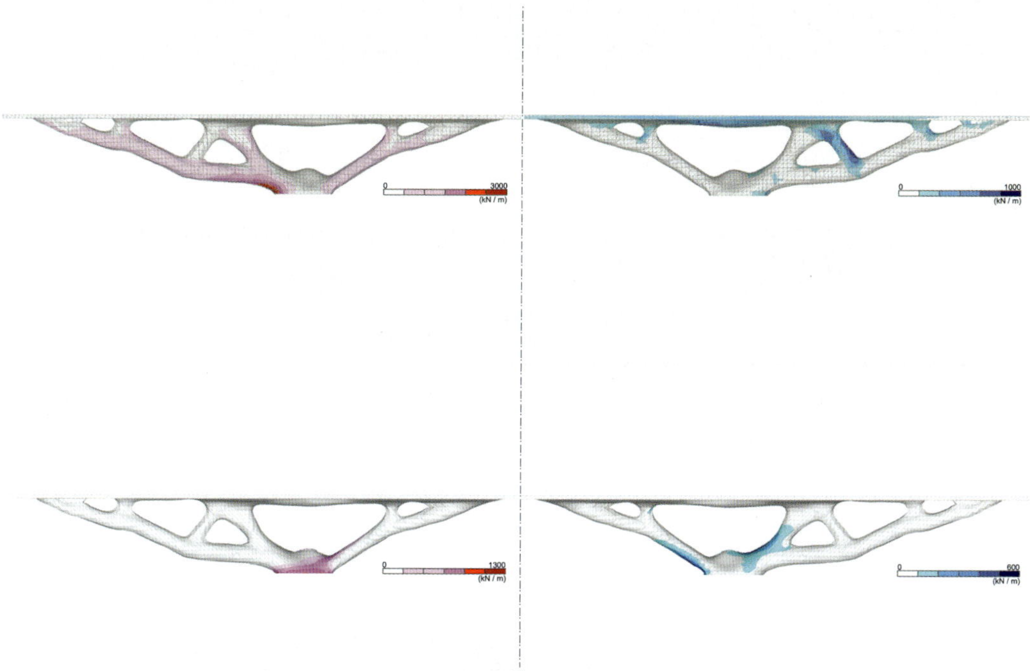

FEM analysis of quarter segment of full model
Top: Distribution of compression
and tension in vertical load
Bottom: Distribution of compression
and tension in thermal load

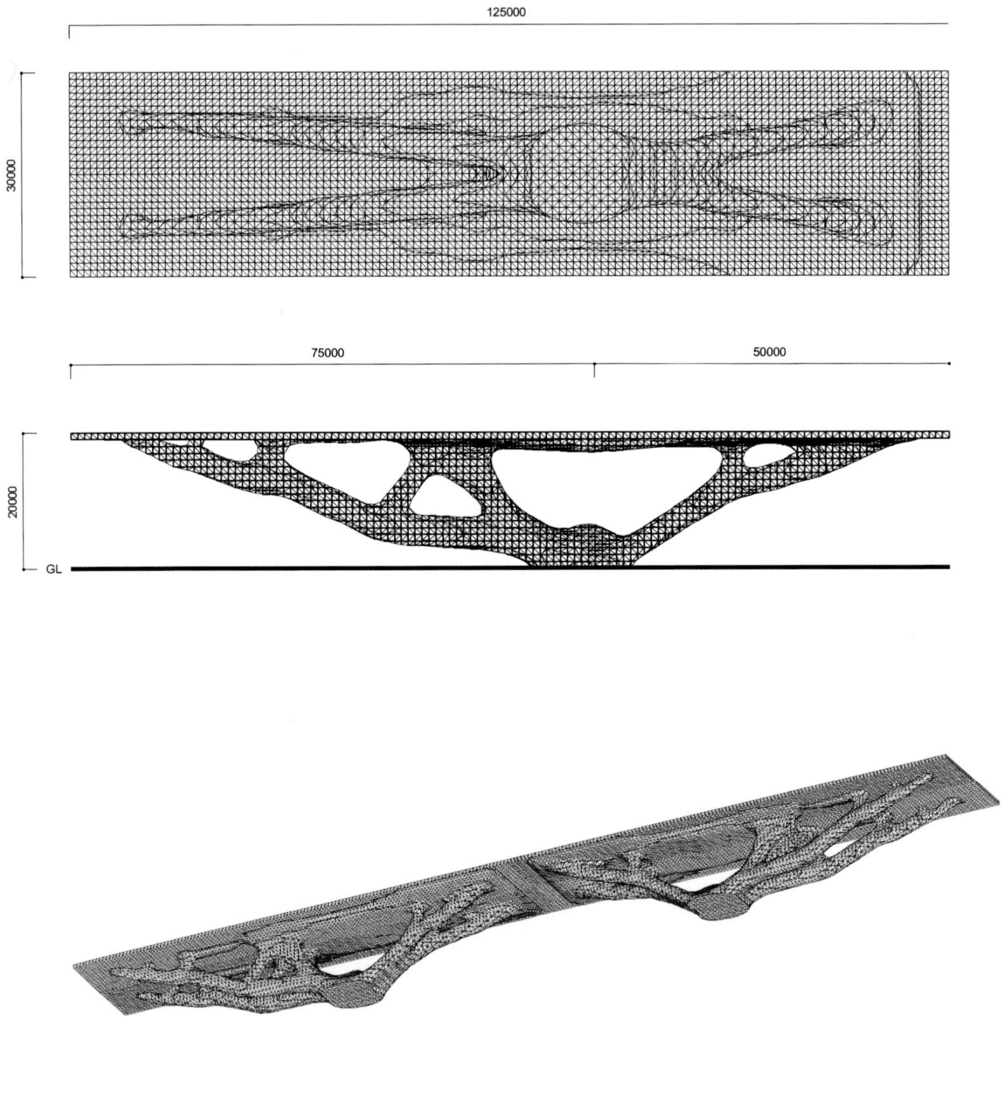

Top: Plan
Centre: Elevation
Bottom: Perspective

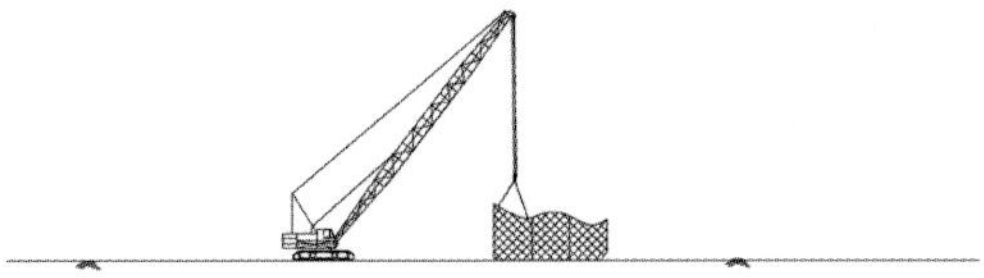

1

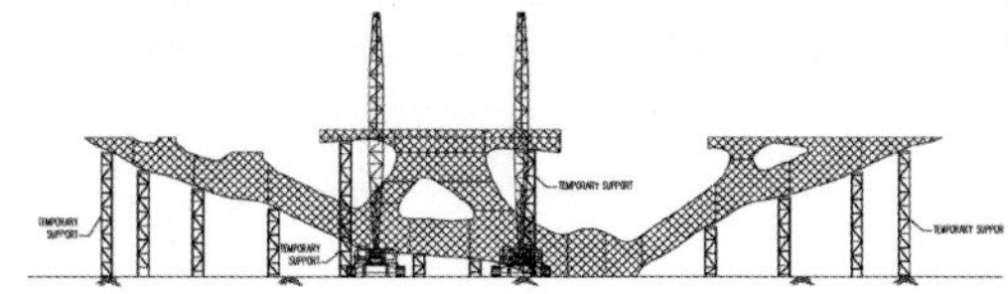

4

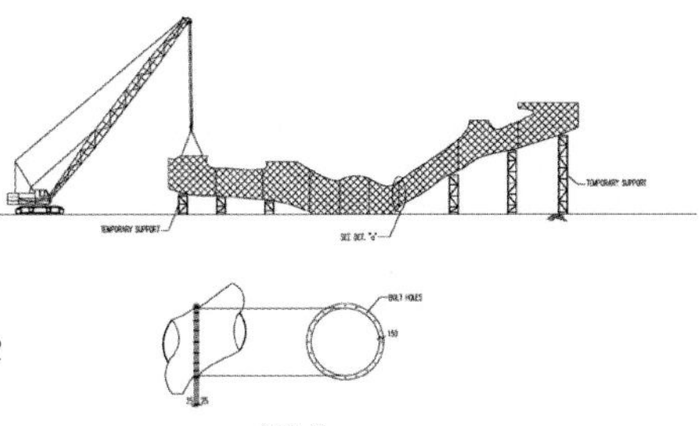

2

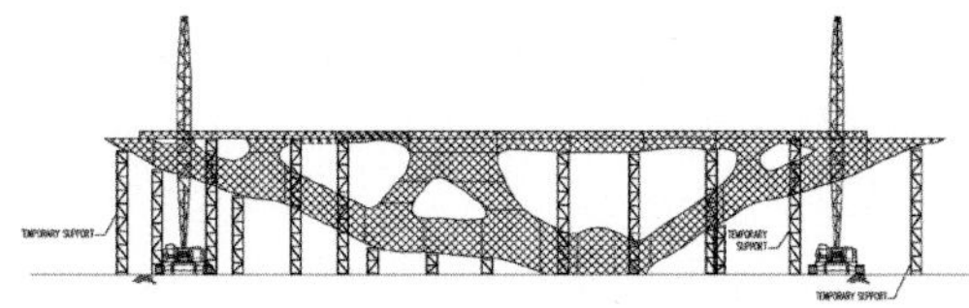

5

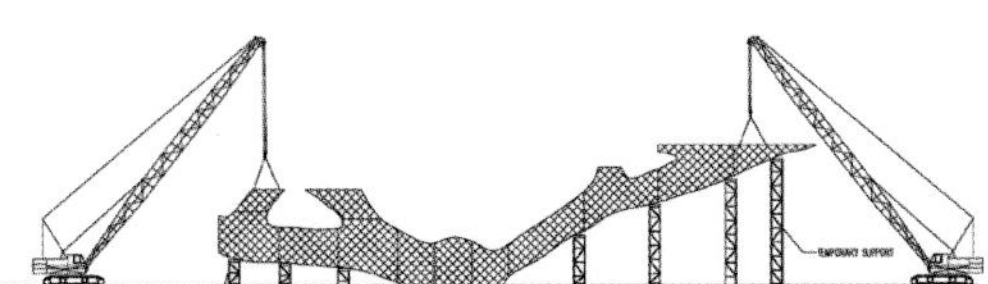

3

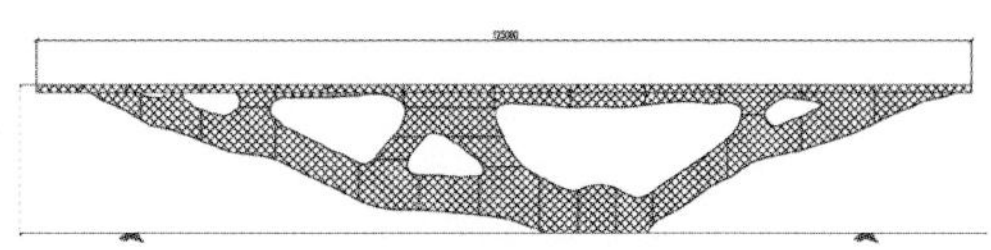

6

Sequence of Assembly

1. Trunk portion is erected by crawler crane
2. Temporary supports are installed and branches erected
3. Installation of supports and branches continues
4. Temporary supports are installed and leaf erected
5. Installation of supports and branches continues
6. Removal of temporary supports

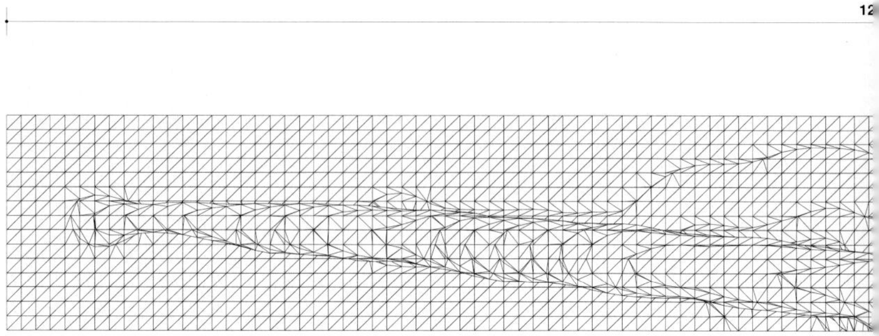

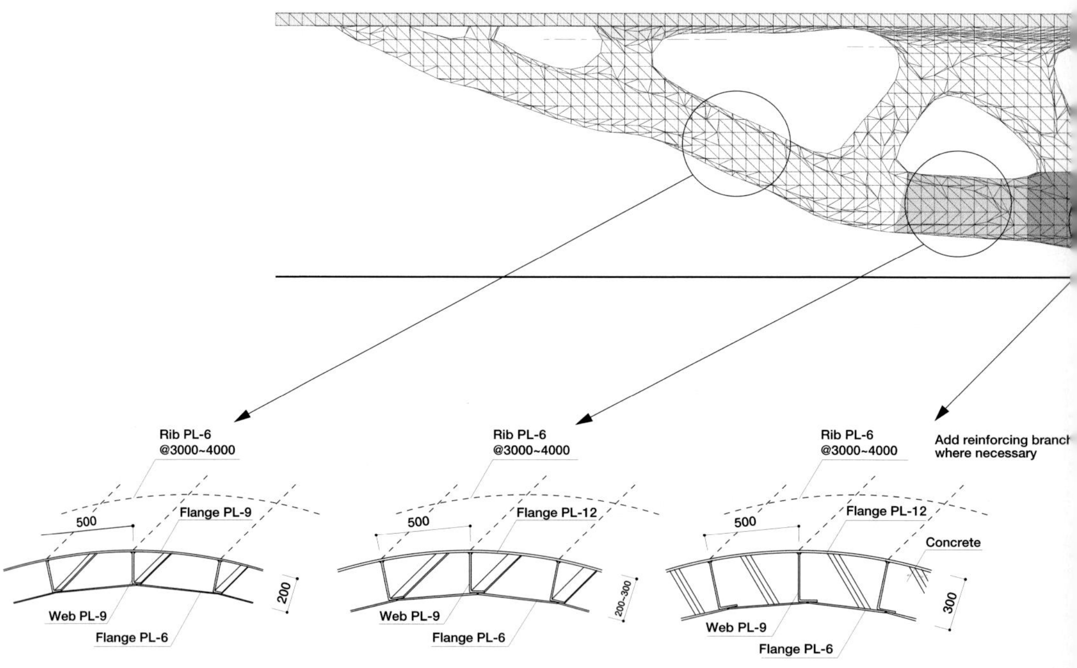

Sandwich Tube A S = 1/30

Sandwich Tube B S = 1/30

Sandwich Tube C S = 1/30

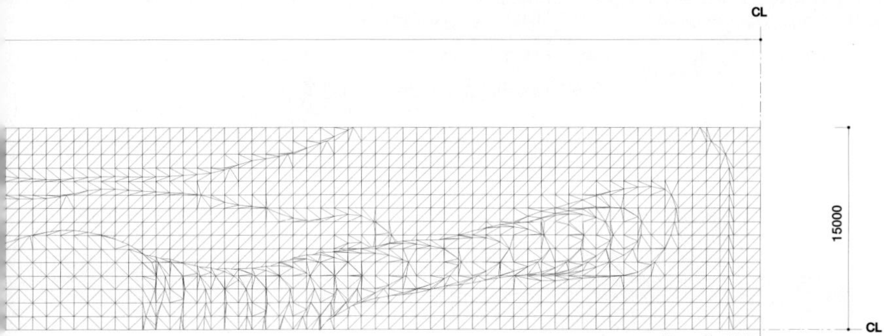

Plan S = 1/400

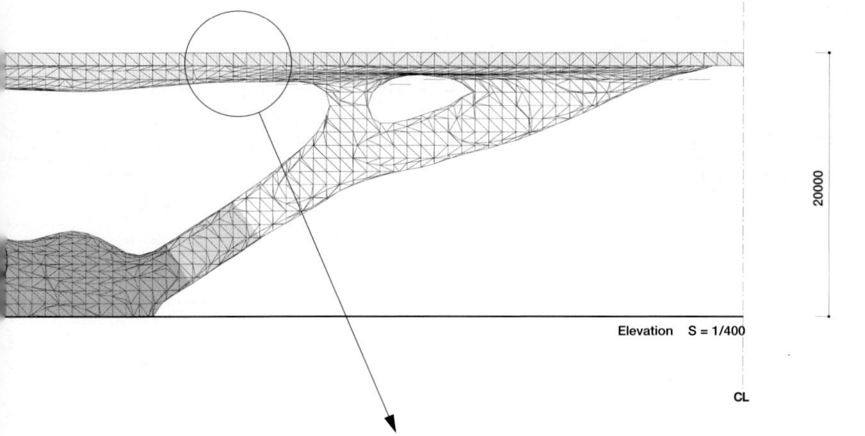

Elevation S = 1/400

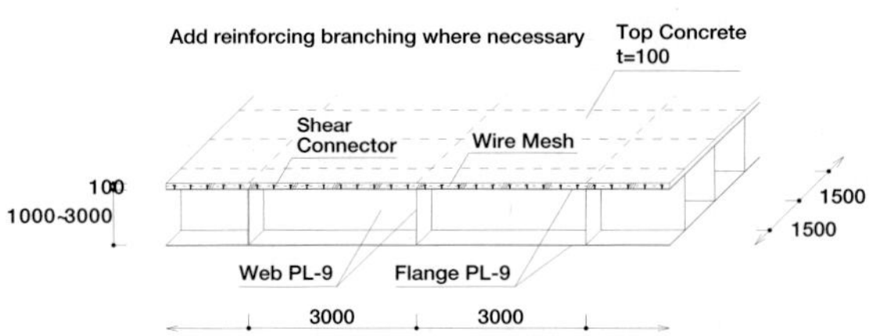

Sandwich Roof S = 1/150

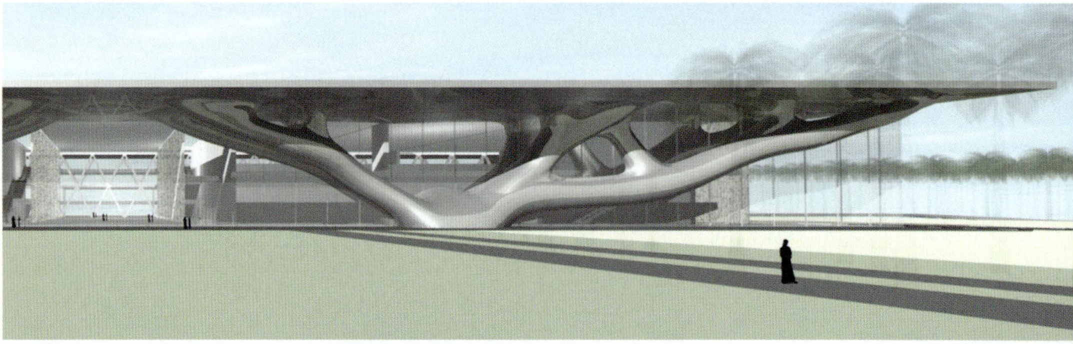

A DIALOGUE WITH TOYO ITO

MUTSURO SASAKI If we trace back to early Modernism the antecedents of contemporary developments in structure, two architects immediately come into view: Mies van der Rohe and Antonio Gaudí. As far as I can see, the impetus for today's original structural forms was provided by these two people, and by Gaudí in particular.

A while ago, in the summer of 1993, I had the opportunity to meet Felix Candela for the first time. As you know, he is the very incarnation of rationalism, a brilliant engineer who has employed extremely thin, lightweight hypar shells in many works of architecture. Candela told me, 'Although he may be my antithesis, the architect I respect above all others is Gaudí.' Up to then all I had apprehended of Gaudí's work was the surface ornamentation. Embarrassingly, I had not noticed its underlying structural importance. Provoked by Candela's remarks, I immediately set off on a pilgrimage around Gaudí's architecture. I reviewed a great deal of documentation and learned about the now-famous hanging model experiments. When I confronted the real thing armed with this knowledge, I was able to perceive Gaudí's ideas and intentions quite clearly. I remember returning home with the strong impression that, even in terms of structure, he was a great architect who comprehended a vast range of issues.

1.

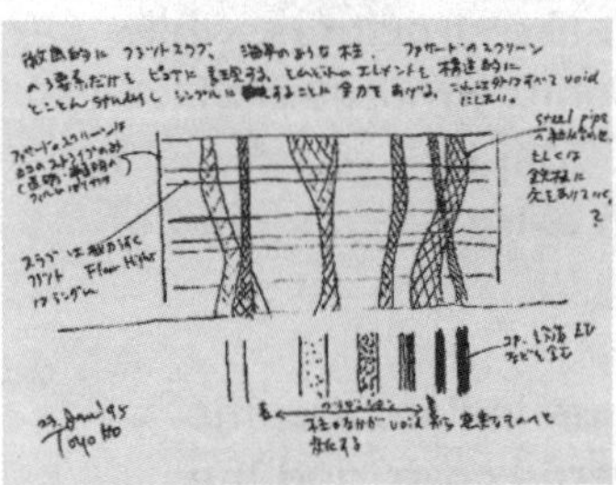

1. Preliminary sketch of Sendai Mediatheque

A year went by, and then you faxed me the first sketch of the Sendai Mediatheque. I can still recall the excitement that welled up inside me the moment I saw it – I intuitively wondered if it might by some chance be posited on an extension of the structures of Mies and Gaudí, although I don't remember exactly how I thought they might have been connected. The floor plates at Sendai stand for Mies's Barcelona Pavilion (1929), with its abstracted image of a floor. The tube including the relation to the plates, on the other hand, was something I found much trickier to work out.

Looking back on that time, I have come to understand that my structural and architectural methodology evolved by drawing on Modernism to some extent, and was by no means discontinuous. This awareness has consistently informed my work since the Sendai Mediatheque, a series of projects that I have now begun to describe as 'flux' structures. I've come increasingly to feel that this kind of architecture, with a structure that flows and melts, is set to become a major methodology, transcending Modernism and contemporary practice.

TOYO ITO It's also clear to me that my own architectural thinking was instantly altered by engaging you to work on Sendai. As you mentioned, the utterly irregular geometries of the tubes were implanted within a homogeneous Miesian space – the 'abstract field' of the floors. This gave rise to various frictions that were tremendously interesting.

I think that the point of Sendai lies in the premise of firstly preparing something homogeneous – notably the thin floors composed of a pattern of ribs – and then catabolising it. During the competition we proposed a regular grid but this immediately began to transform when we entered the basic design phase. In particular, the ribs around the tubes took on a radial arrangement, which meant the moment the tubes were inserted into a homogeneous space you could directly visualise the forces spreading out from them.

Although you have been talking about Mies and Gaudí in general terms, I think they might better be perceived in terms of 'differing geometries'. For the last five years I have been investigating how a relationship might be established between these two geometries. Looking at your work since Sendai, I think it can be roughly divided into two categories. Fukuoka and Kakamigahara are in a sense Gaudí-like structures, while the Kanazawa 21st Century Museum of Contemporary Art (2004) that you did with SANAA embodies an extreme condition of Miesian space.

2.

SASAKI In *Adam's House in Paradise*, the theorist Joseph Rykwert talks of the 'discovered cave' and the 'manufactured tent' as spontaneous architectural original forms. I would paraphrase this and differentiate between two types of structure: 'caves in the earth' and 'houses in the trees'; one returns to the earth, the other takes off from the earth. In other words, for making structures within the planet's field of gravity, there are two readily understandable original forms: 'wanting to return to the earth' and 'wanting to leave the earth'.

3.

The structure of the EPFL Learning Centre (designed by SANAA, scheduled completion 2009) combines the imagery of caves below the earth with floors lifted away from the earth. I thought this integration of the two original forms might be achieved by theoretically expanding my shape design method to include free-curved surface shells that are roofs as well as floors. With no precedents to refer to, however, it became extremely difficult to implement this idea at the developed design stage. Since a shell cannot be established without some degree of rise (the loads must basically be transmitted as an axial force system), many parts were generated with a floor so steep that they became difficult to use. Although there might be various conflicts, I have been focusing all my efforts on devising a consistent structural logic for this project! An illogical and haphazard method of resolution will never achieve any kind of universality.

Mies declared that modern logic is itself homogeneous, and therein lies its universality. He created the concept of a 'universal' space, an open, homogeneous space made of glass and steel. On the other hand, Gaudí tried to find the most universal structural forms in stone and after more than a decade of experimentation came up with a solution that had not previously existed. Similar to Mies and Gaudí, my consistent stance as a structural designer has been to continue searching, in a logical way, for the presence of universality.

ITO I now have a terribly difficult problem, if I consider it logically. On deciding to depart from Modernism, I had two directions to go in: one towards the 'house in the trees', as with the Tod's Omotesando Building (2004), the other towards 'cave-like' things. But with Mies, the 'house in the trees' lost its distinct shape and became a section sliced out of an infinitely extended space. Sendai could in one sense be interpreted in this way. And yet, it could also be described as a drilling of holes through a single compact volume – in which case we could say it's a 'cave'. Thinking of the tubes of Sendai as 'caves' allowed me to begin to treat the building as something with no 'exterior'. If we'd used the tubes on their own, the result would have been grotesque, but by creating a homogeneous space embedded

4.

within a Miesian space we were spared having to think about the exterior. Viewed in this context, the White U (1976) also becomes a tube space, because I was not conscious of the 'outside' in that project.

Le Corbusier was truly a genius at making shapes, so he even managed to imbue cave-like architecture, such as the chapel

2. 21st Century Museum, Kanazawa
3. EPFL Learning Centre, Lausanne
4. Notre Dame du Haut, Ronchamp

at Ronchamp (1955), with an overwhelming sculptural power. My skills are not on the same level, so I always have a problem forming the exterior of the cave-like space. As a hole excavated below the ground, a natural cave doesn't have an exterior.

At Fukuoka the cave-like space, equivalent to the twisted tube, is in itself interesting. However, the moment this appears above ground, it materialises as a shape. And what should be done with this shape?

5.

SASAKI Fukuoka can perhaps be thought of as a dynamically twisted cave with a 'Baroque' shell structure. At Kakamigahara, although we were using the same shape design method of 'Sensitivity Analysis' to search for a free-form surface shape, the result is slightly better organised in architectural terms, more graceful and delicate – akin to a Rococo shell structure, if anything.

As architectural expressions, I have the impression that your work has moved from a contemporary Baroque towards Rococo, but has this been a conscious shift for you? And what do you think about the flow of styles in European architectural history, from the Renaissance to the Baroque, Rococo and then Modernism?

6.

ITO With regard to Kakamigahara, the criteria were originally more architectural than they were with Fukuoka, so they could be articulated into elements that are recognisable as columns and roofs. The work gains a sense of security from these elements, and can also be said to end with 'architecture' in the process.

In moving from Renaissance to Baroque, to Rococo and then to Modernism, I think we have come full circle and arrived back at Primitivism. In other words, wasn't it all over with the Rococo, when everything dissolved?

Similarly, in the modern era, the coining of the phrase 'homogeneous space' led me to question: 'how can homogeneous space be dissolved?'

SASAKI It is not as though everything was dissolved with the Rococo. The Neoclassicism that arose later influenced the ideas of Mies, while the Gothic Revival influenced Gaudí. In other words, there were

7. 

links with early twentieth-century Modernism. Now, at the beginning of the twenty-first century, a similar situation has arisen. However, in the shift from an industrial society to an information society, the speed of change seems to be accelerating, giving the impression that present-day architecture is cycling in an ever-shortening loop.

ITO I think what has come to the fore now is the question of how Miesian space may be distorted. There is another unseen space, possibly called information space, that freely produces distorted spaces in our consciousness. Although everyone is seeking this space of distorted consciousness in architecture, it isn't easy to materialise physically. This limitation is evident in current architecture.

SASAKI When trying to physically materialise this idea, I take the approach of introducing an inherent structural logic, that is to say, mechanical principles. As a student I researched Riemannian space, which is not found in Euclidean geometry but is embedded in N dimensions, similar to very abstract twisted surfaces – that is to say, manifolds. I had some original ideas about how to verify this approach. For instance, employing the tensor analysis used by Albert Einstein, I found that when the excessive areas were abstracted, the resulting twisted surfaces gradually followed conventional shell theory. However, I speculated that under normal conditions something more 'limp' might be embedded in N dimensions. The conclusions of my master's thesis on Riemannian space and the theory of generalised continuum mechanics are now perhaps beginning to appear in the real world.

ITO Listening to what you've just said, I feel that you are thinking about new spatial concepts while somehow imagining very Japanese-style spaces. The transition of European styles from Renaissance to Baroque and Rococo was both a process of dissolving things that had clear shapes and a process of metamorphosis. However, I feel that we Japanese consider spatial concepts within a framework that always relativises pre-existing shapes.

My own architecture in particular has been greatly affected by the 'space of my own inner dialogue'. In this concept, as in the concept of the 'space of sound', there is no exterior.

I think that the condition you're describing – a distorted

5. Island City Central Park, Fukuoka
6. Crematorium in Kakamigahara
7. Church of Colònia Güell, Barcelona

space defined by Riemannian geometry – is extremely beautiful. When it comes to translating this conceptual space into a physical space, however, I can no longer judge what should be done. Embedding this condition in a Mies-like grid might be an easy solution, but I don't want to do that... The big problem is then how to preserve the sense of floating in a space with no exterior.

SASAKI Enthralled by your sketch of Sendai, I found the methodology for realising it gradually came into view. But as you have mentioned previously, a more recent methodology might have suggested different answers for the exterior, departing from the image of a tube embedded in a homogeneous space.

8.

ITO Many interesting states are imaginable simply by asking, 'what would happen if the tubes were made of steel plates?' Speaking of steel plates, when Mikimoto Ginza 2 (2005) was first presented, there was a very intense discussion of the question, 'What exactly is this tense, crisp condition in Mutsuro Sasaki's work?'

SASAKI Thank you for noticing (laughter).

ITO This image of 'crispness' is of a world without thickness that is unresponsive to gravity, a world without depth, which I think is reflected in your unique aesthetic. I have the feeling that where a normal person would use a dimension of ten centimetres, you would say 'let's do it with five!' This is a declaration of a strong will, which leads to a crisp expression.

SASAKI In any project, I have a kind of abstract image in my head from the outset.

ITO Your structures always return to this 'abstract space', related to thinness and delicacy. When you execute a Miesian idea, above all, I feel it has a pristine appearance. Although you say you like Gaudí, I still think your essence is Miesian. When I am in a homogeneous space, I feel a peace of mind. I wonder if this might be close to your own sensibility?

SASAKI In truth, I want to create environments that are touched by the charm of organic nature. This is now at an initial stage of dealing with essentially primitive elements by using free-curved surface shells. In realising such ideals, as well as 'strength as an engineer', I think I am simultaneously required to have 'ethics and aesthetics as an engineer'.

ITO It used to be that on the one hand we had Miesian things, and on the other hand Gaudíesque things. But I think these no longer form a dualistic, opposing schema. It would be utterly dull if we were just to decide that Mies was no good and looked for something more organic as a contrast.

What I am trying to imagine now is a kind of heterogeneity that exists above homogeneity. That is what I am trying to imply by the 'distortion of homogeneity'. Looked at socially, I think the general emphasis on homogeneity tends to lull people into a kind of catatonic state. My own stance is thoroughly opposed to this.

SASAKI I remember Arata Isozaki telling me a while ago that local rules provided the basis for solving local problems. He was looking for states in which the accumulation of these local criteria made the whole.

When I discuss local rules by relating them to homogenisation, I'm the type of person who would rather trust things that are not homogeneous.

ITO I completely agree.

SASAKI If flat floor plates are established at the outset, this becomes linked to the notion of a homogeneous grid division, leading to standardisation and normalisation. It is a fact that there is an a priori desire for horizontal floors in places where people live, but I was sure there had to be other ways of thinking – other kinds of floors. While studying the mechanics of Riemannian geometry I realised I could reach the ordinary theory of plates if the amounts of distortion were all abstracted in mathematical formulas. Because I studied this while I was young, I think heterogeneous states have been implanted in my head from the beginning. Walking on an undulating roof-like surface, as at Fukuoka, or a partially lifting floor, as at EPFL, allows people to reclaim a basic biological instinct: namely, taking care when walking. Contemporary architecture has moved towards extreme homogenisation and optimisation, creating an excess of convenience that has weakened modern people physically and mentally. If a sloping floor was universal, I would

8. Mikimoto Ginza 2, Tokyo

expect human perceptual instincts to be aroused in a different way. Even in this respect, I think it is more normal for the world to be distorted.

ITO Humans originally lived within nature and, as we know, nature itself is heterogeneous. For that reason, locality can be induced. When we generate architecture for a particular location it produces a kind of ordering in which it is generally preferable for the floors to be flat. However, this idea has now been extended to the point where people think that the whole world should somehow be flat, or a place 'where the wind doesn't blow'...

SASAKI This becomes the premise. But in fact, we are getting our priorities confused if we say that.

ITO There has been a total reversal, and it's no longer surprising to hear things like 'wind is abnormal'. If this goes any further, we will find ourselves in a world where there are no differences between people, where 'you and I are the same thing'.

SASAKI As a movement opposed to nature, urbanisation is concerned with how to make an environment that is easy for people to comprehend and control. Among contemporary architects, you have become the leading figure in asking, 'does it really have to be like this?'

My approach, because of my background in Riemannian geometry, is rather, 'isn't distortion more real?' Now, thinking about why I found Sendai so exciting, I believe it was because it triggered this kind of subconscious response in me.

In doing the series of 'flux' structures, I have also wanted to think about architecture and structure more in terms of human instincts. As a structural engineer, however, I'm not qualified to establish architectural concepts and criteria. At best, I can hope to discover things in sympathy with these basic concepts, which will help me develop my ideas further. In this way I would like gradually to shift from a structural model to an architectural model, by which I mean an architecture closer to the natural world. Glimpses of this appear in the free-curved surface shells of Fukuoka and Kakamigahara.

ITO Form in nature is regulated by really simple rules, but these act in complex ways.

SASAKI The interrelationships between elements become extensive, even in the comparatively straightforward field I operate in – mechanics based mainly on gravity. You can understand simple form-giving processes by observing the growth of plants, which take on different shapes depending on the kind of environment they are in. For instance, in a windy place a plant will evolve towards a bending-resistant structural form. In this respect, even taking a single plant as an example, shape determinants can be seen as a mixture of various conditions.

A similar thing occurs in architecture. I think that design consists of compiling these multiple conditions into some shape. The point is to compile these shapes in such a way that their form-giving principles are adequately explained. I find this process interesting. Of all contemporary architects, I think you might be the one to propose answers to these questions. In that sense, I have high hopes for our future collaborations.

This is an edited version of a dialogue first published in Japanese in GA JAPAN 79, March/April 2006. It forms part of a series of texts by Mutsuro Sasaki gathered under the rubric 'Prologues to Contemporary Structure'.

PROJECTS WITH TOYO ITO

ISLAND CITY CENTRAL PARK, GRINGRIN

2005

This greenhouse facility in Gringrin, Fukuoka is 185 metres long by 50 metres wide and composed of a single continuous surface twisting twice to form three blocks. Each block is made of RC free-curved surface shell (400 millimetres thick) and spans 50 metres by partially touching the ground. The shape design (Sensitivity Analysis) method was used to modify the initial shape imagined by Toyo Ito for the main roof. This generated the optimum structural shape with the least possible bending stress and a minimum of strain energy and deformation.

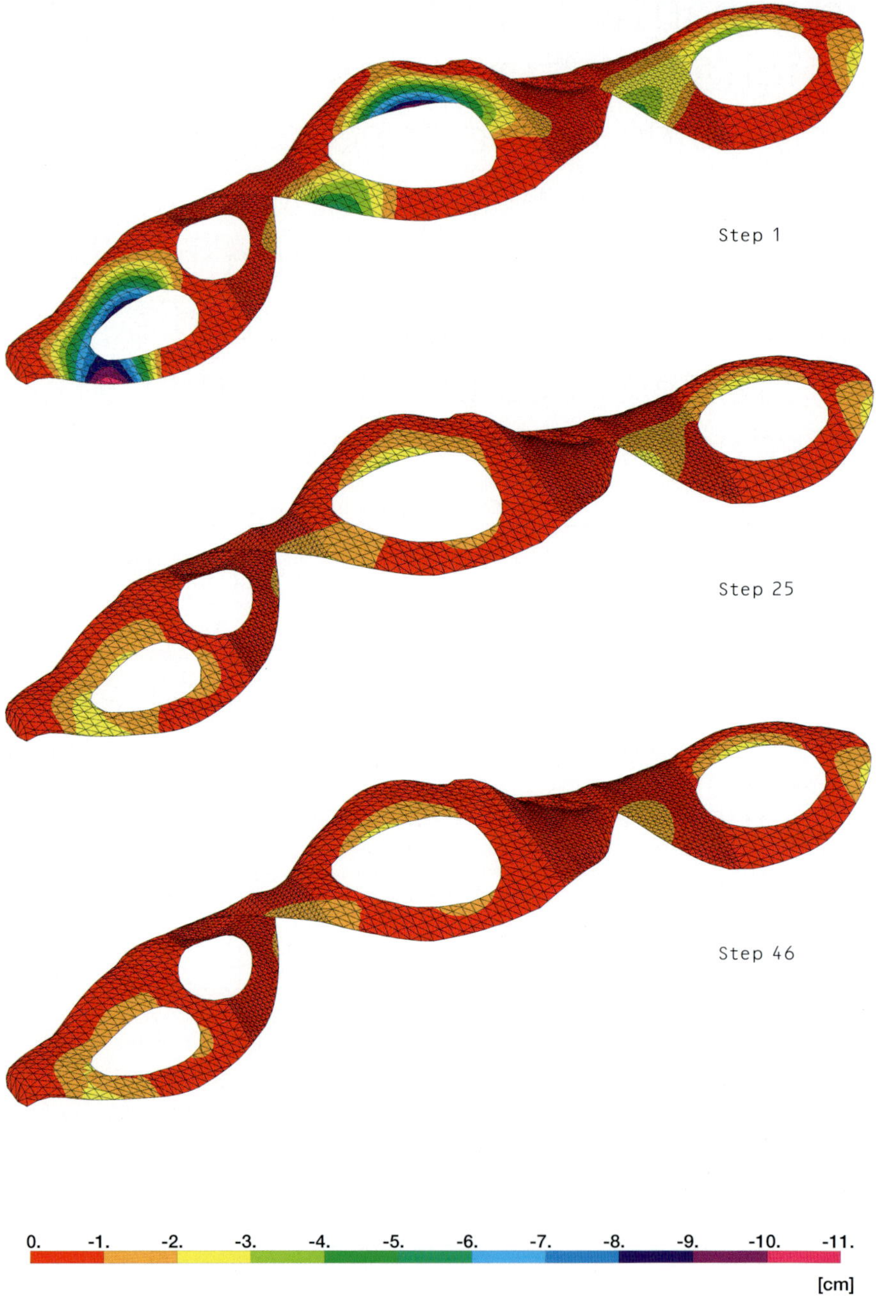

Shape design by Sensitivity Analysis
The figures above show the sequence of the shape analysis from step 1 (the initial condition) to step 46 in relation to the deformation at the roof by a uniformly distributed vertical load (1.5t / m2). The overall deformation is reduced as the shape evolves.

Formwork for the shell
Arrangement of steel rods

Concrete pouring
Removal of supports

General view during construction

General view

Interior view

PROJECTS WITH TOYO ITO

CREMATORIUM IN KAKAMIGAHARA

2006

This is a design for a crematorium in Gifu, 80 metres long by 60 metres wide. The bearing walls and columns are arranged with a span of about 20 metres and support the main roof, which is made of RC free-curved surface shell {200 millimetres thick}. The shape design {Sensitivity Analysis} method was used to modify the initial shape imagined by Toyo Ito for the main roof. This generated the optimum structural shape with the least possible bending stress and a minimum of strain energy and deformation.

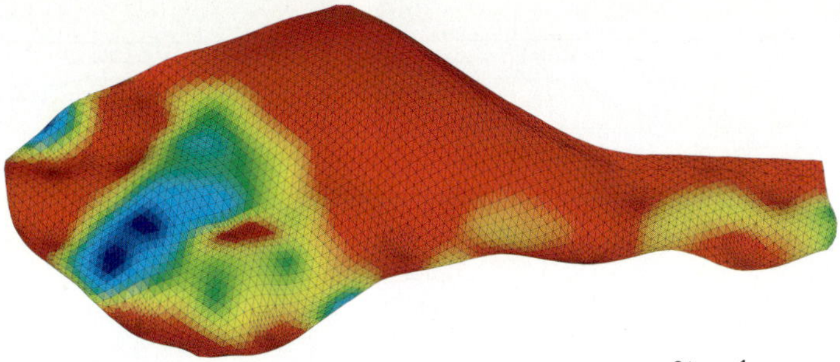

Step 1

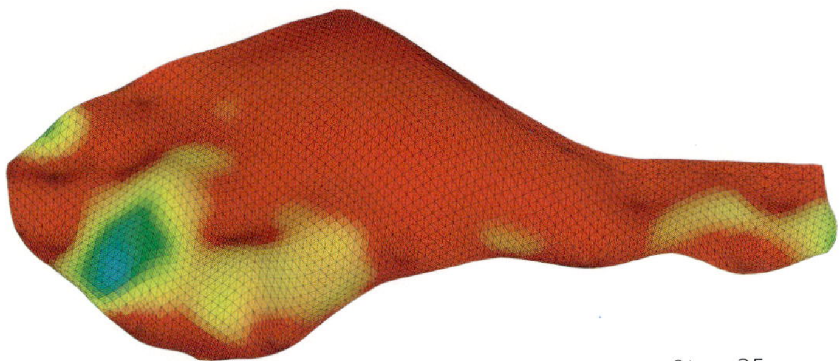

Step 25

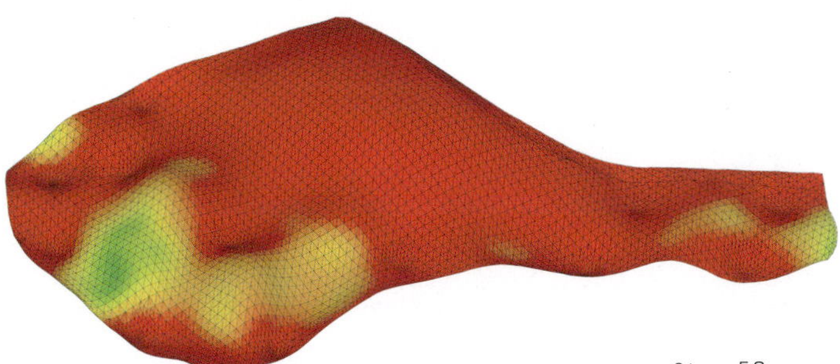

Step 50

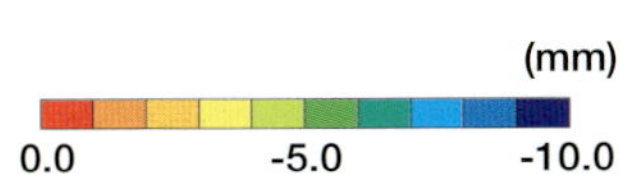

Construction of formwork and supports
for concrete shell

Above: Curved surface of form at top of column

Arrangement of steel rods for the shell
Concrete pouring

Jack down process
Removal of supports

Construction sequence

THEORY: SHAPE DESIGN BY SENSITIVITY ANALYSIS AND THE 3D EXTENDED ESO METHOD

SHAPE DESIGN BY SENSITIVITY ANALYSIS METHODS TO DETERMINE A FREE-CURVED SURFACE

EXISTING METHODS

A method involving the hanging model experiment used by Gaudí and Isler to define a funicular surface (a type of free-curved surface). A method involving the iterative local modification of the curved surface (imagined by an architect then described by spline function) according to its condition of stress and deformation analysed by FEM.

PROPOSAL FOR A NEW METHOD

Shape design method using Sensitivity Analysis to create a rational free-curved surface by the mechanical theory that minimises strain energy in a structure.

SENSITIVITY ANALYSIS: FORMULATION

Design Parameter: Shape (z Coordinates)

Evaluation Function: Strain Energy C

Focusing on the transition of strain energy during transformation dz at node No. i in a structure composed by 'n' elements.

$$C = \frac{1}{2}\{f\}^T\{u\}$$

$$[K]\{u\} = \{f\}$$

$$\alpha_i = \frac{dC}{dz_i} = -\frac{1}{2}\{u\}^T \frac{d(\sum_e [K_e^{(i)}])}{dz_i}\{u\}$$

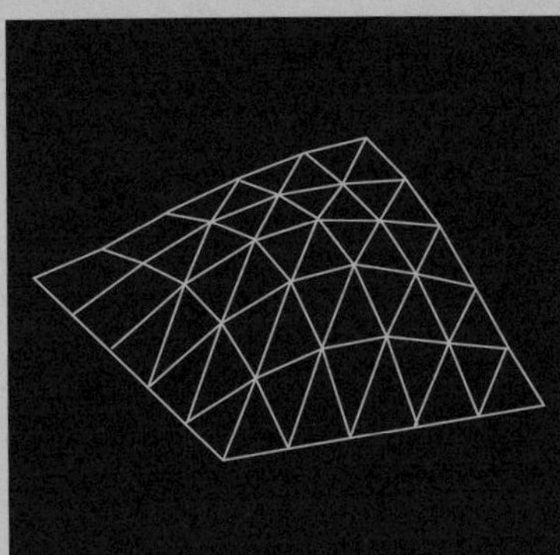

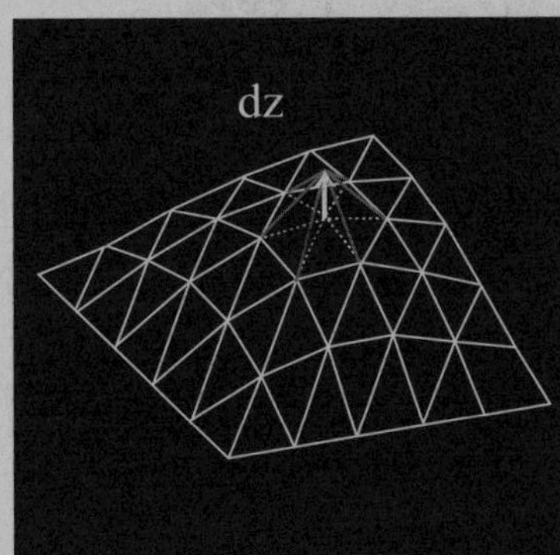

$\{f\}$ Nodal Load Vector

$\{u\}$ Nodal Displacement Vector

$[K]$ Element Stiffness Matrix

$\sum_e [K_e^{(i)}]$ Sum of Element Stiffness Matrix relating to node No. i

Evaluating how much a transformation at a node influences the strain energy's transition in the whole structure.

MODIFICATION OF SHAPE

$\alpha_i > 0$

in case of increases in the strain energy by the transformation at node 'i' → Pull down its z coordinate

$\alpha_i < 0$

in case of decreases in the strain energy by the transformation at node 'i' → Push up its z coordinate

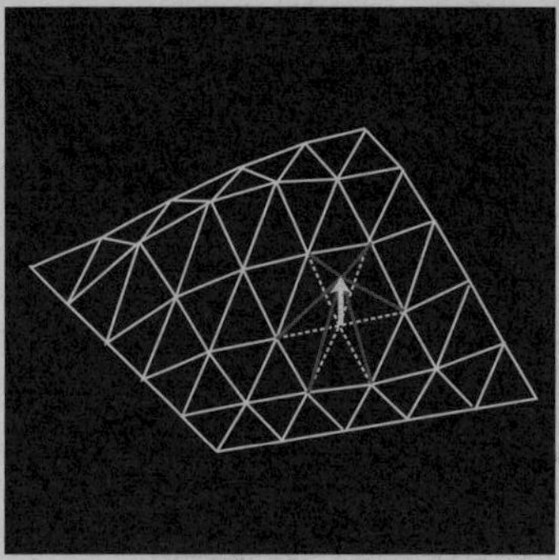

$$\rightarrow \alpha_i$$

$$z_i' = z_i - \alpha_i \Delta z$$

z_i' = z coordinate after modification

z_i = z coordinate before modification

Δz = Parameter for adjusting modification amount

ENDING CONDITIONS OF EVOLUTION

$C_k - C_{k-1} \rightarrow 0$

(Few changes in strain energy during a new evolving step)

$\delta_{max} \leq \delta_0$

(Maximum deformation falls below allowable deformation)

ANALYSING THE ALGORITHM

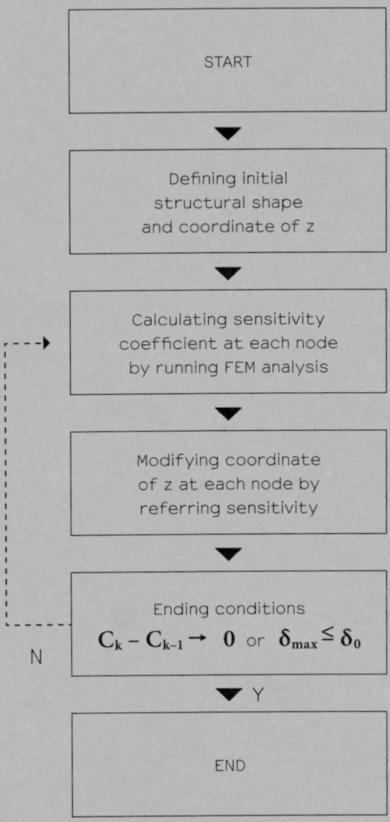

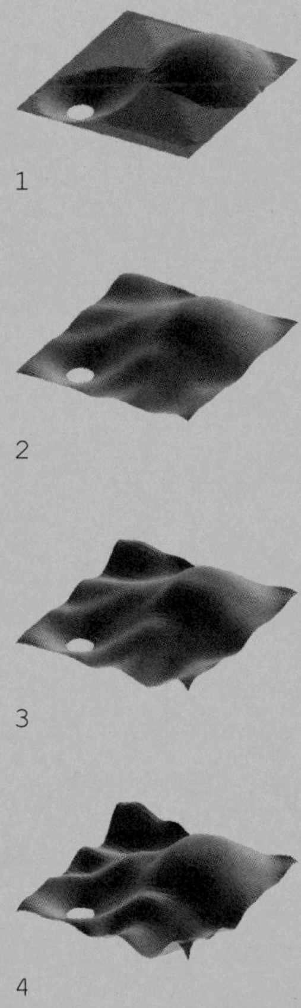

Case study

1. Defining initial structural shape and z coordinate (design parameter)
2. Calculating sensitivity coefficient at each node by running FEM analysis
3. Modifying z coordinate at each node by referring to sensitivity
4. Repeating operations 2–3 till ending conditions of evolution are reached

In general, transmission efficiency occurring during bending is extremely low in comparison with axial forces such as tension and compression. In other words, a structure that possesses even distributions among all structural members without bending moments has the highest transmission efficiency, allowing it to be fabricated with the minimum amount of materials. This is the ideal goal of shape resistance in spatial structures, where structural optimisation can be utilised for the shape design. The shape of a structure is closely related to its mechanical behaviour. Change in a structure's mechanical behaviour simultaneously corresponds to a modification of its shape.

The 3D Extended Evolutionary Structural Optimisation (ESO) method can generate a rational structural shape by organically apprehending the close relation of shape and forces in the whole structure.

The method can seek a rational three-dimensional structure from a mutual evolution process by not only deleting inefficient elements step by step but also by multiplying elements.

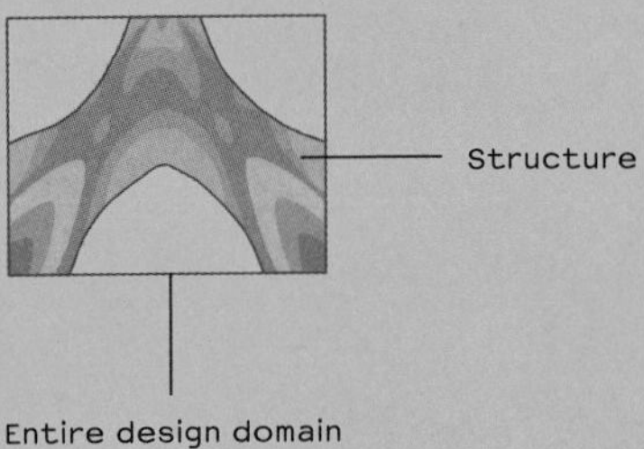

1. Introduction of contour lines
2. Deletion and multiplication
3. Operation by distribution of Von Mises' stress in structure
4. Evaluation of structural shape

$$\sigma^{\text{Mises}} = \frac{1}{2} \sqrt{(\sigma_x - \sigma_y)^2 + (\sigma_y - \sigma_z)^2 + (\sigma_z - \sigma_x)^2 + 6(\tau^2_{xy} + \tau^2_{yz} + \tau^2_{zx})}$$

ORDINARY ESO METHOD =
EVOLUTION BY DELETION

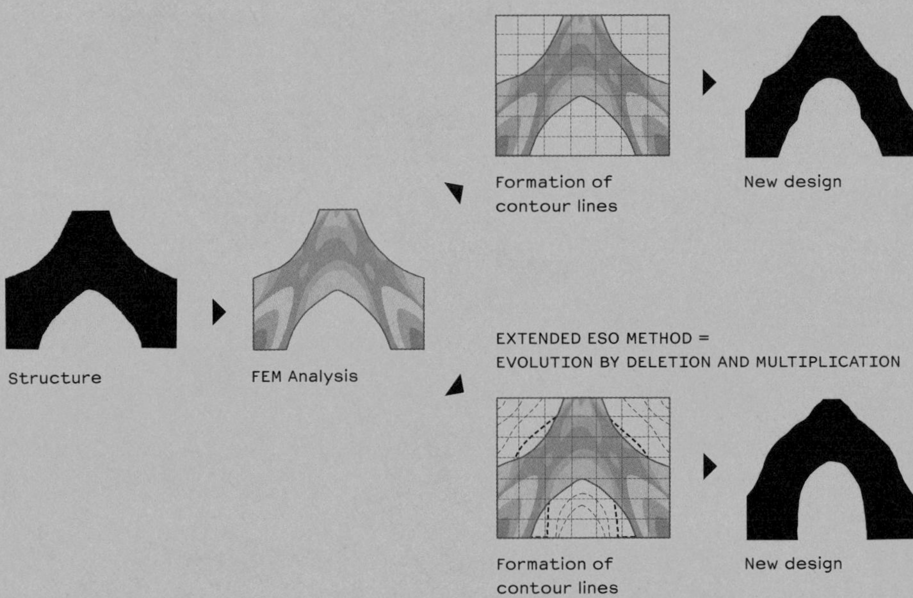

----- $\sigma < \sigma_c$
----- $\sigma = \sigma_c$
—— $\sigma > \sigma_c$

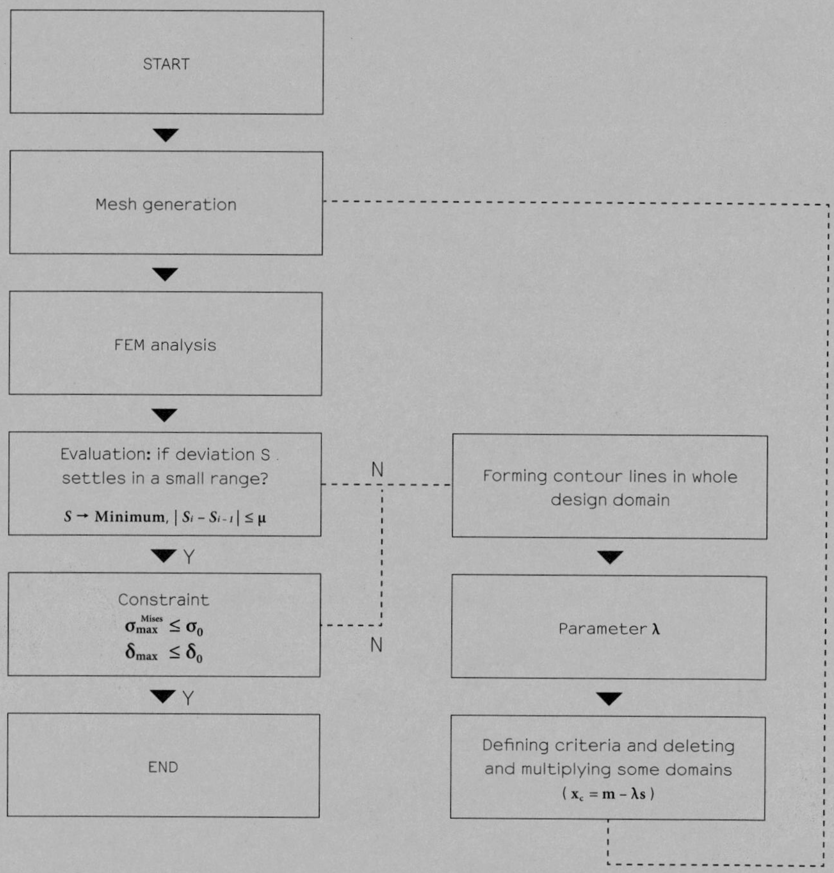

m = mean
s = deviation
x_c = criteria

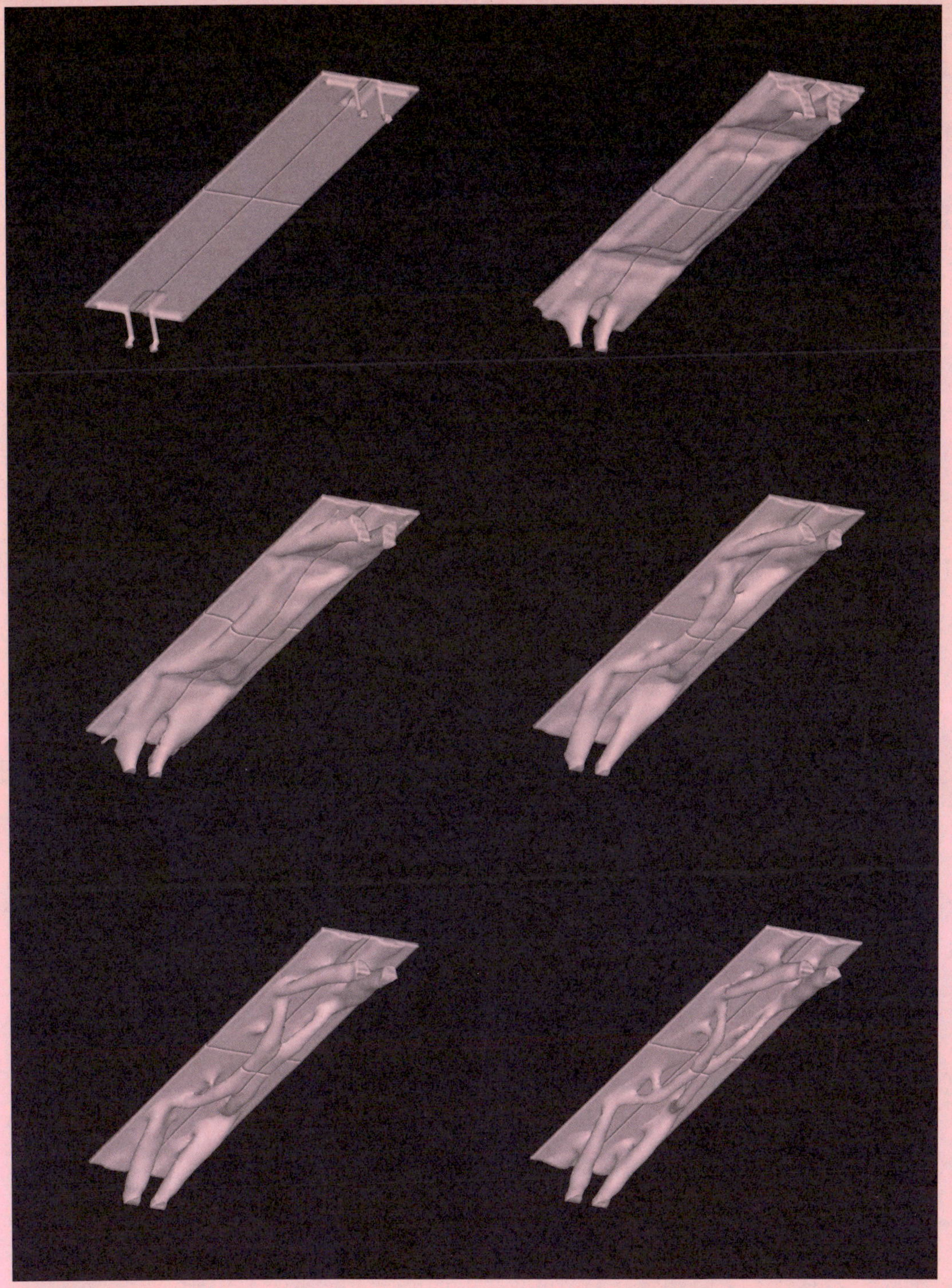

Case study

MUTSURO SASAKI
STRUCTURAL DESIGNER

BIOGRAPHY

1946	Born in Aichi, Japan
1968	BA Architecture, Nagoya University
1970	MArch, Nagoya University
1970-79	Kimura Structural Engineers, Engineer
1980	Sasaki Structural Consultants, Founder
1998	Doctorate in Engineering, Nagoya University
1999-2004	Professor at Nagoya University
2004-	Professor at Hosei University

MEMBERSHIP OF INSTITUTIONS

AIJ	Architectural Institute of Japan
JSCA	Japan Structural Consultants Association
JIA	The Japan Institute of Architects
IASS	International Association for Shell and Spatial Structures
IABSE	International Association for Bridge and Structural Engineering

MAJOR AWARDS

1991	Matsui Gengo Prize: Miwa Lock factory, Tamaki
2002	BCS Prize: Sendai Mediatheque
2002	JSSC Award (achievement): Sapporo Dome
2003	Prize of AIJ: Sendai Mediatheque
2004	BCS Prize: Louis Vuitton Omotesando
2004	Tsuboi Prize, IASS: Computational Morphogenesis of 3D Structures by Extended ESO Method
2006	BCS Prize: Kanazawa 21st Century Museum
2006	Good Design Award: Mikimoto Ginza 2
2006	JSSC Award (Thesis): Guideline for Redundancy of Spatial Structure

RECENT MAJOR WORKS

2000	*Sendai Mediatheque* Toyo Ito, Miyagi, Japan
2001	*Sapporo Dome* Hiroshi Hara, Hokkaido, Japan
2002	*Louis Vuitton Omotesando* Jun Aoki, Tokyo, Japan
2003	*Yamaguchi Information Art Centre* Arata Isozaki, Yamaguchi, Japan
2003	*Christian Dior Omotesando* Kazuyo Sejima+Ryue Nishizawa, Tokyo, Japan

2004 *Matsumoto-City Civic Hall* Toyo Ito, Nagano, Japan
2004 *Kanazawa 21st Century Museum* Kazuyo Sejima+Ryue Nishizawa, Ishikawa, Japan
2005 *Kitagata Community Centre* Arata Isozaki, Gifu, Japan
2005 *Island City Central Park Gringrin* Toyo Ito, Fukuoka, Japan
2005 *Mikimoto Ginza 2* Toyo Ito, Tokyo, Japan
2006 *Crematorium in Kakamigahara* Toyo Ito, Gifu, Japan
2006 *Hospital Cognacq-Jay* Toyo Ito, Paris, France
2006 *Glass Pavilion, Toledo Museum of Art* Kazuyo Sejima+Ryue Nishizawa, Toledo, Ohio, USA
2006 *Novartis Campus* Kazuyo Sejima+Ryue Nishizawa, Basel, Switzerland
2006 *Design School Zollverein*, Kazuyo Sejima+Ryue Nishizawa, Essen, Germany
2007- *Stadstheater Almere* Kazuyo Sejima+Ryue Nishizawa, Almere, The Netherlands
Qatar Education City Convention Centre Arata Isozaki, Doha, Qatar
Ivam extension Kazuyo Sejima+Ryue Nishizawa, Valencia, Spain
New Museum of Contemporary Art Kazuyo Sejima+Ryue Nishizawa, New York, USA
Montjuic 2 Toyo Ito, Barcelona, Spain
EPFL Learning Centre Kazuyo Sejima+Ryue Nishizawa, Lausanne, Switzerland
Louvre-Lens Kazuyo Sejima+Ryue Nishizawa, Lens, France
Redevelopment of Fiera Milano Historical Fairground Arata Isozaki, Milan, Italy